# HEALTH EFFECTS OF AMBIENT AIR POLLUTION

# HEALTH EFFECTS OF AMBIENT AIR POLLUTION

## How safe is the air we breathe?

Jane Q. Koenig
University of Washington.
U.S.A.

Kluwer Academic Publishers
Boston/Dordrecht/London

**Distributors for North, Central and South America:**
Kluwer Academic Publishers
101 Philip Drive
Assinippi Park
Norwell, Massachusetts 02061 USA
Telephone (781) 871-6600
Fax (781) 871-6528
E-Mail <kluwer@wkap.com>

**Distributors for all other countries:**
Kluwer Academic Publishers Group
Distribution Centre
Post Office Box 322
3300 AH Dordrecht, THE NETHERLANDS
Telephone 31 78 6392 392
Fax 31 78 6546 474
E-Mail <services@wkap.nl>

Electronic Services <http://www.wkap.nl>

---

**Library of Congress Cataloging-in-Publication Data**

Koenig, Jane Q.
Health effects of ambient air pollution : how safe is the air we breathe? / Jane Q. Koenig
p. cm.
Includes bibliographical references and index.
ISBN 0-7923-7719-2 (alk.paper)
1. Air--Pollution--Health aspects. 2. Asthma--Environmental aspects.
I. Title.

RA576.K525 1999
615.9'02--dc21 99-048297

---



*Printed on acid-free paper.*

Printed in the United States of America

To Marci and Nathan

Contents

ABBREVIATIONS

| | |
|---|---|
| AM | Alveolar macrophage |
| BAL | Bronchioalveolar lavage |
| BHR | Bronchial hyperresponsiveness |
| CAA | Clean Air Act |
| CFCs | Chlorofluorocarbons |
| CO | Carbon monoxide |
| COHb | Carboxyhemoglobin |
| COPD | Chronic obstructive pulmonary disease |
| EIB | Exercise-induced bronchospasm |
| EPA | Environmental Protection Agency |
| ETS | Environmental tobacco smoke |
| $FEF_{25-75}$ | Forced expiratory flow in the mid portion of a FVC maneuver, from 25-75% |
| $FEV_1$ | Forced expiratory volume in one second |
| FVC | Forced vital capacity |
| HAPs | Hazardous air pollutants |
| $H^+$ | Hydrogen ion |
| $H_2SO_4$ | Sulfuric acid |
| ICAM | Intercellular adhesion molecule-1 |
| IgE | Immunoglobulin E |
| IL | Interleukin; there are many forms -1 through -13 |
| INF $\gamma$ | Interferon gamma |
| LDH | Lactose dehydrogenase |
| NAAQS | National Ambient Air Quality Standards |
| $NO_2$ | Nitrogen dioxide |
| PAF | Platelet activating factor |
| PFTs | Pulmonary function tests |
| PM | Particulate matter |
| $PM_{10}$ | Particulate matter less than or equal to 10 micrometers |
| $PM_{CF}$ | Particulate matter between the sizes of 2.5 and 10 micrometers |
| $PM_{2.5}$ | Particulate matter less than or equal to 2.6 micrometers |
| Pb | Lead |
| ppb | Parts per billion |
| ppm | Parts per million |
| Raw | Resistance of the airways |

| | |
|---|---|
| Raw | Specific resistance, Raw multiplied by the lung volume at the time of the measure |
| $SO_2$ | Sulfur dioxide |
| TLV | Total lung volume |
| TNF-α | Tumor necrosis factor, alpha |
| TSP | Total suspended particulate matter |
| UV | Ultraviolet light |
| VOCs | Volatile organic compounds |
| $\mu g/m^3$ | Microgram per cubic meter |

List of Figures

List of Tables

Acknowledgements

Several of my colleagues read various chapters of this book and made invaluable comments. I thank Brook Madrone, Dan Jaffe, Therese Mar, Gary Norris, Dan Luchtel, and Paul V Williams for their careful reading of specific chapters. Therese Mar also helped me with many of the graphs. I thank Nathan Koenig for assistance with indexing. I also thank the students in my class on the health effects of air pollution, Spring quarter, 1999 who read the chapters for class and pointed out errors and areas to be clarified. Finally I express my deep gratitude to William E Pierson who taught me a lot about asthma and inspired our early studies into the health effects of air pollutants.

# CHAPTER 1.

# INTRODUCTION: *SCOPE OF THE BOOK*

Welcome Sulphur Dioxide
Hello, Carbon Monoxide
The air, the air
is everywhere

Breathe deep
While you sleep
Breathe deep

(cough)
Deep
(cough, cough)

FROM: Hair, the musical: Ragni and Rado, 1969

This book was written to fill a gap. I teach an upper division class/graduate course on the Respiratory Health Effects of Air Pollution in the Department of Environmental Health at the University of Washington. I would like to require a text, but have not been able to find one suitable for the goals of my course. The result is this book which is designed for undergraduate upper class students or graduate students. Some knowledge of biology and chemistry will help the reader get more out of this book. However, that is not a prerequisite. The book is written as if it were to be used as a textbook, however, I hope that it will be useful to a general audience interested in the topic of the health effects of ambient air pollution.

As the world approaches a new millennium, citizens are faced with various problems caused by human population and activities. Three of these are involved with waste products of civilization: air pollution, water pollution, and soil pollution. Most human activities in the year 2000 require the consumption of energy: burning coal to generate electricity, burning oil or natural gas to operate factories, burning gasoline or diesel fuel to power mobile fleets or car, buses, and light-duty or heavy-duty trucks. The price citizens pay for the use of these products to support their life styles is increased amounts of pollution. The theme of this textbook is air pollution: sources, health effects, regulation, and future needs.

One of the reasons that air pollution is associated with a wide array of respiratory health effects is that inhalation is the primary route for entry of toxic agents into the body. We have very little ability to choose the air that we breathe. If the water from the faucet appears brown or smells, one can choose to acquire drinking water elsewhere. However, when we step outside or enter a building, and the air does not smell good, we still have to continue breathing. Lung disease is common; One in five persons in US have a form of chronic lung disease. Occupational lung disease is the most common occupational disease.

To set the stage for reading this book, several points need to be emphasized.

1. More than 50% of current outdoor ambient air pollution is due to mobile source emissions. When you look at polluted air in large cities in the US, Europe, and other do not automatically blame industrial smoke stacks alone. Also blame yourself if you drive a car or light duty track. If you ride a bus, the bus is contributing to air pollution although the ratio of pollution per individual is much more favorable than the single occupancy vehicle (the most prevalent situation). Major sources of air pollution in the western states are various forms of vegetative burning, either wood burning for residential heat or field burning for agricultural purposes.

2. A second point this book documents is that current air pollution levels in the US and Europe (and Asian and South American cities have even higher concentrations of most pollutants) are associated with adverse health effects, even to the point of premature mortality.

3. A third point the reader should note is that everyone is not affected equally by air pollution. There is a wide range of susceptibilities among individuals, which are just now beginning to be understood. Infants, children, and older persons with comprised health appear to be the most sensitive. There are suggestive data that show gender differences in response to one common air pollutant, ozone. Also a given individual may be more or less susceptible at certain times depending on co-pollutants and personal host factors (a recent upper or lower respiratory tract infection).

The text is organized into several sections. First there are brief summaries of both the structure and function of the respiratory system. However the reader should seek out appropriate texts if detailed information is desired. Also, although the respiratory system has been considered traditionally to be the target organ of air pollution effects, it is now apparent that some air pollution (particulate matter) has an adverse effect on the cardiovascular system as well. What are the types of adverse effects one sees following exposure to common air pollutants? Changes in lung capacity, changes in airway hyperresponsiveness, inflammation of the bronchial airways causing edema and swelling of the airways and consequent obstruction of air flow and chest tightness and wheezing, irritation of either the lower or upper respiratory system causing cough or nasal congestion, proliferation of mucus causing production of phlegm, coagulation of blood, and cardiac dysrhythmias all have been associated with various air pollutants.

There are three basic fields of scientific research that contribute to our knowledge of the health effects of air pollution. These are animal and in vitro toxicology, controlled human laboratory research, and epidemiology. The fields of animal and in vitro toxicology have provided a great deal of research designed to unravel the dose-response relationships between lung structure, function, and biochemistry and various air pollution concentrations. Toxicology also has sought to identify biologic mechanisms underlying these effects. No matter how elegant and carefully designed the toxicologic studies are, we are still left with the ultimate question: What are the effects of life-time exposure to air pollution for a given population or a given individual? Controlled laboratory studies have determined precise relationships between air pollutant concentrations and lung function decrements and release of inflammatory mediators in the nose or bronchial airways. And yet, due to the short-term nature of controlled human studies, we are again left with the same question: What are the effects of lifetime exposure to air pollution for a given population or a given individual? The field of epidemiology is devoted to examining associations between certain conditions of human behavior and inter-individual characteristics (diet, exercise levels, body weight, smoking history, occupational exposures, gender, ethnicity) and subsequent health outcomes for years. Thus this field is the appropriate field to pose the question we have in mind; What are the effects of lifetime exposure to air pollution for a given population or a given individual? However in order to answer the question, the epidemiologist must know the true life time (or shorter term) exposure to each air pollutant for each individual, and whether the exposure(s) occurred during infancy, during exercise, alone or in a mixture, at an similar daily average or in peaks and troughs. This need for better information on real life exposures to air pollution is discussed in the chapter on exposure assessment.

A separate chapter is devoted to asthma since asthma may be the disease syndrome most likely to increase an individual's risk of an adverse health effect from air pollution. Asthma is the most common chronic childhood disease. The prevalence of asthma and asthma mortality are both on the rise. Since there is a proven relationship between asthma aggravation and air pollution concentrations, asthma will be used in this text as a model for the study of adverse effects of air pollution. Some argue that since asthma rates are increasing at a time when concentrations of major air pollutants are decreasing, air pollution cannot be responsible for the perplexing increase in asthma. This interpretation requires considerable confidence in our knowledge of what is the true profile of urban air. Although the levels of several of the most common air pollutants (sulfur dioxide, carbon monoxide, $PM_{10}$) have decreased since 1970, we obviously do not know the concentrations and trends for the air pollutants which are not required to be measured daily. Moreover, the size of the mobile fleet is continuously growing.

Descriptions of documented health effects of the individual criteria air pollutants (sulfur oxides, particulate matter, ozone, nitrogen dioxide, and carbon monoxide) fill the majority of the pages in the text. After reading these chapters, the reader should be able to differentiate among the typical health effects of each of these pollutants.

Although outdoor pollutants demand most of the attention of the US EPA, it is generally agreed that indoor air pollution is a major problem for the majority of the population. It may be that our "home is our castle", but there is no bill of rights entitling us to clean indoor air. One chapter is devoted to a description of the sources of indoor air pollution and the health effects that are associated with indoor environments.

Up until 1997, most of the health concern regarding use of fossil fuels dealt with outdoor air pollutants. In addition to the climate change predicted by the International Panel on Climate Change, there now is concern for the health effects due to global warming. The chapter on climate change attempts to tie together the parallel issues of the health effects of air pollution and the potential health effects of global warming.

The field of risk assessment devoted its embryonic years to projecting risks of increases in cancer for agents assumed to be carcinogens. Currently, there is increasing interest in developing guidelines for calculating risk for non-cancer endpoints. The long and comprehensive process EPA uses to develop a criteria document to support the National Ambient Air Quality Standards for each of the criteria pollutants is a form of risk assessment as described in the concluding chapters. The final chapter is a listing of some of the most important research needs at the end of the millennium.

# CHAPTER 2.

# STRUCTURE OF THE RESPIRATORY SYSTEM

One of the reasons that air pollution is such a threat to human health it that humans have no choice over the air that they breathe. Thus in our homes, outdoors, and in workplaces, we often breathe air which is not as clean as we would prefer. The average adult breathes approximately 10,000 to 12,000 liters of air each day of life. Individuals who exercise frequently or have jobs that involve strenuous activity breathe a much greater volume of air. Inhalation is the major route of entry into the body for toxic materials. Given the volume of air inhaled, it is not surprising that respiratory illnesses are the fourth most common illness or that occupational lung disease is the most common occupational disease. In order to prepare the reader for discussions of air pollutant-respiratory system interactions, this chapter will give a brief summary of the structure of the respiratory system. The structure of the respiratory system largely determines its function. It is recommended that the reader consult a respiratory system textbook if more detail is desired.

Figure 2-1 is a schematic of the respiratory system. The respiratory system is divided into the upper respiratory system including all structures before the trachea (numbers 1 through 8 in the schematic) and the lower respiratory system which includes the airways from the trachea to the alveolar spaces. Important structural components of the human respiratory system are given in Table 2-1.

*Table 2-1. Components of the human respiratory system.*

*Nose and nasal passages*
*Thorax*
*Muscles of respiration*
- *Intercostals: Internal and External*
- *Diaphragm*
- *Abdominals*
- *Accessory*

*Lungs*
- *Right Lobe: Upper--Middle--Lower*

*Left Lobe: Upper--Cardiac Notch--Lower*
*Airways*
*Larynx*
*Trachea*
*Bronchi*
*Right and left mainstem bronchi*
*Bronchioles (varying smooth muscle: cartilage ratios)*
*Conducting*
*Terminal*
*Respiratory*
*Alveolar ducts, alveolar sacs, alveoli*
*parenchyma, acinis, capillaries*
*Cells*
*Epithelial cells (lining the airways);*
*Type I pneumocytes ;Type II pneumocytes ; Goblet and Serous cells;*
*Ciliated cells; Mast cells; Alveolar macrophages*
*Nervous innervation*
*Brain stem (motor nucleus of cranial nerve X (Vagus)*
*Autonomic nervous system:*
*Beta adrenergic innervation: Airway dilation*
*Muscarinic cholinergic innervation: Airway contraction*

---

Type I pneumocytes form the gas exchange barrier; Type II pneumocytes secrete surfactant; goblet and serous cells secrete mucus); ciliated cells move the mucociliary escalator; mast cells secrete various mediators of inflammation; and alveolar macrophages engulf, deactivate and remove foreign substances from the lung.

For the purposes of understanding the health effects of air pollution, it is educational to consider how various structural components of the respiratory system interact with inhaled gases and particles. In general, concern regarding the adverse effect of air pollution on the respiratory system has centered on the lower respiratory system. However, as will be described in succeeding chapters, the upper respiratory system also is a target for adverse effects from air pollution and will be discussed first.

## UPPER RESPIRATORY SYSTEM

The upper respiratory system is made up of the nares and the nasal turbinates. Often the nose is ignored as part of the respiratory system, however, most of air intake enters through the nose. Nasal breathing is the route of choice during quiet activities for all individuals without some type of nasal obstruction. Even during exercise when we switch to oral breathing approximately 30-35% of the air is inhaled through the nose. The nasal passages play a very important role in respiration by warming, humidifying, and cleansing the air we breathe. Thus the nose acts as a first line of defense for the lungs. Many large particles deposit in the nasal passages, and thus are prevented from entering the lungs. Also water soluble gases are largely scrubbed from inspired air by the nose. For instance almost 99% of low concentrations of $SO_2$ inhaled during quiet breathing is taken up in the nasal passages. Formaldehyde, a common industrial contaminant, also is extremely water soluble and most likely taken up to a great degree in the nasal passages. Consequently, the target organ for adverse effects from formaldehyde is the nose; increased cases of nasal carcinoma have been reported in individuals exposed to formaldehyde.

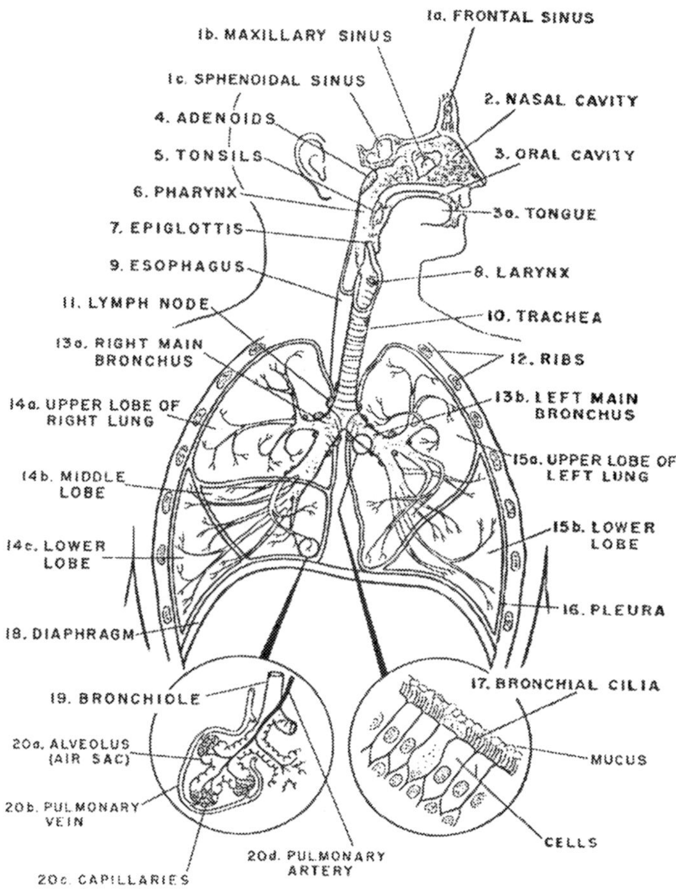


*Figure 2-1. The respiratory sytsem. FROM: American Lung association with permission.*

Figure 2-2 is a schematic of a cross section through a human nose. The legend identifies the epithelial cells lining the nasal passages.

Important cells located in the nose are mucus secreting cells, ciliated cells for removing particles which land on the nasal surfaces, and mast cells release inflammatory mediators. Note the large surface area that comes in contact with gases and particles that are inhaled through the nose.

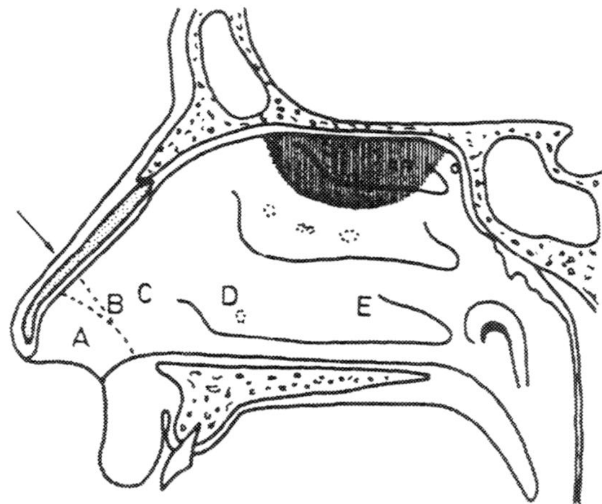


*Figure 2-2. Lateral wall of the nasal cavity. A ) skin in the nostril; B) squamous epithelium without microvilli; C) transitional epithelium with short microvilli; D) Pseudostratified columnar epithelium with few cilia; E) Pseudostratified columnar epithelium with many cilia. FROM: Proctor, 1982.*

## LOWER RESPIRATORY SYSTEM

The respiratory system has three basic functions: 1) ventilation, 2) gas exchange with the blood, and 3) respiration --the exchange of gases at the cellular or tissue level. Ventilation involves the airways. Gas exchange occurs at the gas-blood interface in the alveolar space. Actual respiration (delivery of oxygen and uptake of carbon dioxide) occurs at the tissue level throughout the body. As can be seen, there is extensive branching to the airway. Refer back to Table 2-1 to follow the descriptions of the various components of the lower respiratory system.

### The Bony Thorax

The thoracic cavity actually defines the size of the lungs. During development the lungs tend to grow until this growth is restricted by the thoracic wall. The rib cage provides protection for the lungs. The ribs are associated with the internal and external intercostal muscles.

## Muscles of Ventilation

Muscles of inspiration pull the thoracic cavity down and outward. These muscles are the diaphragm, the external intercostals, and occasionally the accessory muscles. The diaphragm is dome shaped in its relaxed position but is pulled flatter during inspiration. Inspiration is an active process initiated by the central nervous system. As the thoracic cavity is enlarged during inspiration, the lung size increases and lung pressure decreases. This provides a pressure gradient for air to enter the lungs. Expiration on the other hand, often is a passive process. When inspiration ceases, elastic properties of the lungs provide an elastic recoil which forces the lungs to resume their resting size. This elastic recoil pushes air out of the lungs, and the ventilatory cycle is then ready to repeat. However, expiration can be forced. In fact, when the capacity and function of the lungs are measured, the measurements usually take place during forced expiration. Thus, the two most common pulmonary function measurements are the forced expiratory volume in one second ($FEV_1$) and the forced vital capacity (FVC). By their very nature, these measurements require subject cooperation. Muscles of expiration are the abdominal muscles and the internal intercostals, as well as the relaxation of the diaphragm.

## The Lungs

As stated earlier, the lungs are divided into lobes; three lobes to the right lung and two lobes to the left (see figure 2-3). The top of the lungs, or apex, extends slightly above the first rib. The bottom, or base, of the lungs is in contact with the diaphragm. The indentation of the left lung, where the left middle lobe has been displaced by the heart, is referred to as the lingula or cardiac notch.

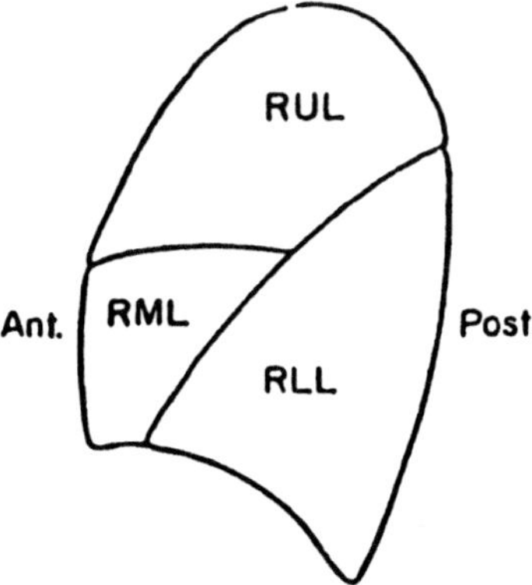


*Figure 2-3. A schematic of the lobes of the right lung. (Note: The left lung has no middle lobe due to displacement of that space by the he*

**The Airways—General Description**

Figure 2-4 shows an extremely simplified schematic of the branching of the airways in a human lung according to Weibel, the father of lung morphometry.

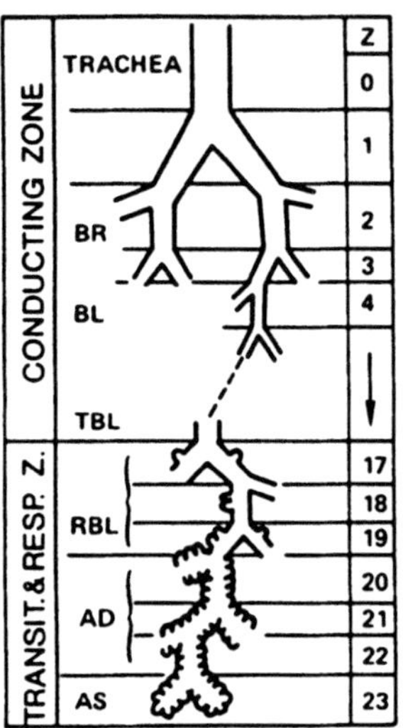


*Figure 2-4. A simplified schematic of the generations of the human airways. FROM: West, 1995 with permission.*

In this schematic there are 23 generations or branchings of the airways. Dissecting the airways from a human lung is extremely tedious and only a few lungs have been dissected. The number of generations varies and there is probably considerable inter-individual variation in the numbers of generations. Note that there are two airway zones depicted, a conducting zone and a transitional/respiratory zone. The first 16 or so divisions of the airways are termed the conducting airways. The denotation is based on the fact that little or no gas exchange takes place in these airways and, from a functional point of view, these airways simply move air in and out of the lungs. The most peripheral airways are termed at first, transitional airways--transitional from conducting to respiratory--and then true respiratory airways. Their unique structure is described later. The largest airways are termed bronchi that are then followed by bronchioles. The ratio of smooth muscle to cartilage increases from the trachea down to the end of the bronchi. After that point, beginning with the bronchioles, cartilage is not present in the walls of the airways. The smooth muscle of the airways is intertwined with elastic fibers that contribute to elastic recoil.

**a. Larynx**

The larynx is known informally as the voice box. In general, it connects the oral and nasal passages with the trachea. It is composed of smooth muscle and can constrict in a condition known as laryngospasm. This effectively closes off all the airways since it is a single passage.

**b. Trachea**

The trachea is a somewhat rigid tube constructed mainly of cartilage. The lumen of the trachea is held open by incomplete, C-shaped, cartilaginous rings. The posterior membranous portion contains smooth muscle. The trachea bifurcates at the carina into two mainstem bronchi (right and left) which, through multiple branchings, deliver air to the right and left lung. Notice in Figures 2-1 and 2-5 that the right mainstem bronchus is considerably shorter and turns on a more acute angle than the left mainstem bronchus. The right mainstem bronchus supplies air to the right upper lobe, the right middle lobe and the right lower lobe. The left mainstem bronchus supplies air to the left upper lobe and the left lower lobe. The shorter length and steeper angle of the right mainstem bronchus has functional consequences. As particles in inhaled air travel through the lung they tend to impact at the bifurcations of the airways. More particles impact the walls at the right mainstem bronchus bifurcations than at the left due to the sharper angle. The right mainstem bronchus is the more common site of bronchogenic carcinoma.

**c. Bronchi**

The two mainstem bronchi, left and right,in turn give rise to the lobar bronchi that ventilate each of the lobes of the. They divide in a series of asymmetric bi-and tri-furcations into segmental and subsegmental bronchi. The development of the fiberoptic bronchoscope has allowed better knowledge of the order of the dividing bronchi. As mentioned earlier, with each generation the smaller bronchi have less cartilage and more smooth muscle.

**d. Bronchioles**

The next order of airways is the bronchioles that totally lack cartilage. The lumen of the bronchioles range from a few millimeters to less than one millimeter in an inflated lung. Bronchioles are defined as conducting or respiratory airways. The last of the conducting bronchioles prior to the respiratory bronchioles are called terminal bronchioles. Conducting bronchioles primarily serve the function of ventilation; that is, simply passing oxygen down to the gas-exchange area. The conducting bronchioles are referred to as an anatomic dead space since little gas exchange occurs in this area. The volume of air in the dead space is approximately 150 ml. A rule of thumb is that each individual's dead space is approximately equal to their weight in pounds. In contrast the respiratory bronchioles contain air sacs where gas exchange takes place.

**e. Alveolar Ducts, Alveolar Sacs, Alveoli:**

Respiratory bronchioles give rise to fully alveolarized generations termed alveolar ducts. There are approximately 300 million alveoli in a human lung. Each alveolus is approximately 0.3 mm in diameter. The alveolar surface area is estimated to be between 70 and 100 square meters, approximately the size of a tennis court. The unique structure of the alveoli is ideal for the function of gas exchange with the capillaries of the systemic blood system. Figure 2-5 is a diagrammatic representation of the human respiratory unit referred to as the acinus. The acinus contains one respiratory bronchiole and many alveolar ducts and sacs. The interalvelolar septem allows interchange of air between alveoli. The reader should appreciate that this vast surface is ideal for the rapid exchange of gases between the lungs and the vascular system. Nevertheless, due to the tiny diameter of alveoli, they contribute only about 10% to lung volume and total airway resistance.

## Gas-exchange Area

The blood gas barrier is less than 0.5 micrometers in diameter, making it very easy for oxygen to diffuse from the alveolar space into the capillaries of the vascular system and likewise for carbon dioxide to diffuse from the capillaries into the lung and be removed during expiration. Both oxygen and carbon dioxide move across the blood gas barrier by means of simple diffusion along concentration gradients. The percent of oxygen in inhaled air is 21% whereas the concentration in the capillaries entering the lungs is about 16%. On the other hand the percentage of $CO_2$ in inhaled air is less than 1% and the percentage in capillaries entering the lungs (and in exhaled breath) is 5%.

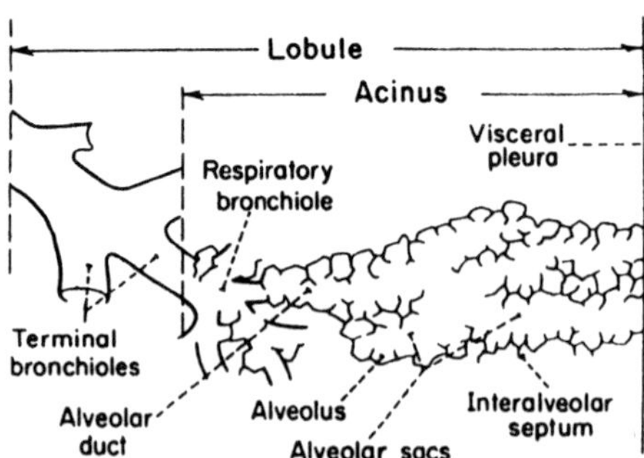


*Figure 2-5. A diagram of an acinus showing the relationship between the terminal bronchioles, the respiratory bronchioles, and the alveoli. FROM; Hildebrandt, 1989 with permission.*

## Cell Types in the Lung

The lungs contain a large number of cells. I will only introduce some of the more numerous which have important functions. Table 2-2 gives a summary of the cell types and their function.

*Table 2-2. Common cell types in the lung.*

| *Type of cell* | *Function* |
|---|---|
| *Goblet or mucous cell* | *Synthesizes and secretes mucus* |
| *Ciliated cell* | *Moves foreign material along the mucociliary escalator* |
| *Basal cell* | *Provides structure* |
| *Kulchitsy cell* | *Contains neurosecretory granules; synthesizes peptides* |
| *Serous cell* | *Contains secretory granules--located centrally* |
| *Clara cell* | *Contains secretory granules--located distally. suggested role in asthma* |
| *Alveolar macrophage* | *Engulfs foreign matter; secretes inactivating enzymes; defense mechanism.* |
| *Pneumocytes* | |
| *Type I* | *Lines the alveolar sacs* |
| *Type II* | *Secretes surfactant* |

## Nervous Innervation

Both branches of the autonomic nervous system play major roles in the function of the lungs. The sympathetic nervous system is primarily represented by beta$_2$-adrenergic receptors in the airways of the lungs. Stimulation of the beta$_2$-adrenergic receptor causes dilation of the airways. This action has been capitalized on by drug companies. The most common, widespread bronchodilating medication used by patients with asthma is a beta$_2$ agonist (examples are albuterol or salbuterol). The parasympathetic nervous system is represented in the lungs by branchings of the vagus nerve (the tenth cranial nerve). Stimulation of the vagus nerve causes bronchoconstriction. Again this information has been incorporated into pharmacologic interventions for asthma. Both atropine and ipratropium bromide are parasympathetic antagonists (drugs that act in an opposite manner) and are used to induce bronchodilation. These medications are not as widely used as the beta$_2$-adrenergic medications. The phrenic nerve is involved in respiration through its innervation of various respiratory muscles.

## Structural Components

Elastic fibers are present at every level of the respiratory system. They allow the lungs and airways to stretch during inspiration so that outside air can flow into the system they provide the elastic recoil that forces air out of the system during expiration. The airways are composed of relative distributions of cartilage, elastic fibers, and smooth muscle, as described earlier. Cartilage is present throughout the bronchi but is absent in the bronchioles. Elastic fibers are represented throughout all the airways.

Due to the smooth muscle lining the airways, many stimuli can trigger contraction of this muscle, a condition called bronchoconstriction or bronchospasm. This is often caused by action of the vagus nerve. Bronchoconstriction, a type of airway obstruction, causes airway narrowing and results in a smaller volume of air being forcibly exhaled. Airway narrowing also can occur as a result of inflammation of the airways that causes edema and swelling. The third common cause of airway narrowing is proliferation and thickening of the mucus lining layer which decreases the functional caliber of the airways.

The structure of the airways is very important in asthma. Lungs of patients with asthma are known for the thickened basement membrane, the presence of excess mucus and even mucus plugs, and chronic inflammation. Airway remodeling in asthma is currently a topic of great research interest. Asthma will be discussed in more detail in Chapter 8.

# CONCLUSION

In summary, this brief overview of the structure of the respiratory system should help readers understand the functional measures relied on for assessing the health effects of air pollutants.

# REFERENCES

Hildebrandt J. Respiration. IN Textbook of physiology. Patton, Fuchs, Hille, Scher, Steiner (eds). WB Saunders, Philadelphia, 1989.

Netter F. The Ciba collection of medical illustrations. Volume 7: Respiratory system. Ciba Pharaceutical Co. Summit, NJ, 1979.

Proctor DF, Andersen I. The nose: Upper airway physiology and the atmospheric environment. Elsevier Biomedical Press. New York, 1982.

West JB. Respiratory physiology—the essentials. Fifth edition. Williams and Wilkins, Baltimore, 1995.

# CHAPTER 3.

# PHYSIOLOGY OF THE RESPIRATORY SYSTEM

The respiratory system includes the upper respiratory system (mainly the nasal passages) and the lower respiratory system (the airways of the lungs). As mentioned earlier, the respiratory system is the primary route of exposure of the body to toxic substances. Respiratory rates vary tremendously by age, gender, size, and activity level. An average adult human inhales approximately 10,000 liters of air per day. We inhale and exhale approximately 500 ml of air with each breath and take about 12 breaths per minute during quite periods. Ventilation increases rapidly with exertion.

## LUNG FUNCTION

Measurements of lung function will be discussed first, followed by a description of nasal function. Lung function values are predictable based on age, gender and height. Charts called nomograms are available that can be determined whether a particular lung function measurement is within normal predicted ranges. Most computerized spirometers calculate the percent of predicted values automatically if the appropriate subject information is entered. Lung function also is quite reproducible and this has made it an extremely useful tool for assessing lung changes in an individual attributable to air pollution exposures. Lung assessments have been made following controlled exposures to specific pollutants in a laboratory setting and also in various settings in the field, either comparing an individual before, during and after an air pollution episode or by comparing lung function in residents across two or more geographical regions.

The most common method for measuring lung function is spirometry, a technique that requires the subject or patient to take a deep breath and blow it out as hard and as fast as possible into the spirometer tubing. Figure 3-1 illustrates the spirometric procedure and shows the lung volumes that can be measured. This maneuver requires the full cooperation of the subject, but is quite straight-forward and most people can accomplish it successfully. Spirometry is used regularly in physicians' offices to measure lung function and is used in the workplace for lung function surveillance to determine whether functional changes are occurring more rapidly than expected. The American Thoracic Society (ATS) has published Guidelines for standardization and approved techniques for performance of spirometry (1994).

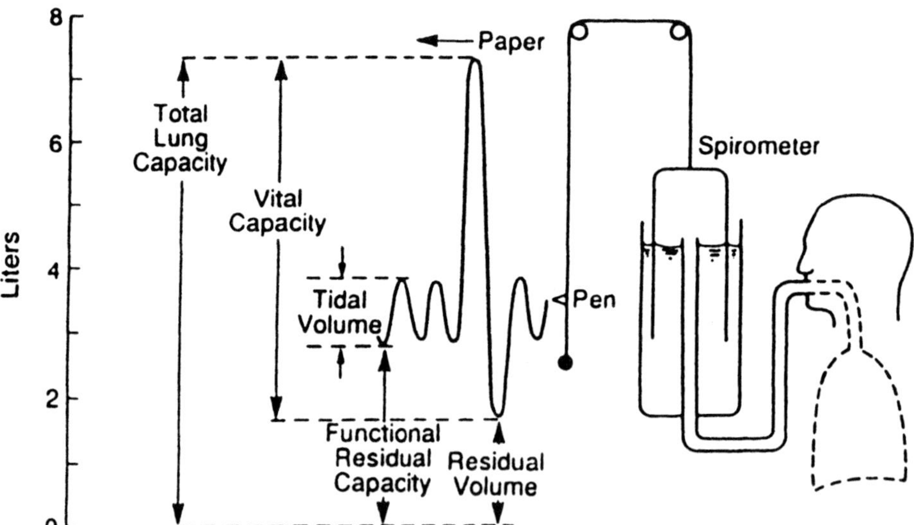


*Figure 3-1. A schematic of the use of a spirometer showing the lung volumes that can be measured. Note that the residual volumes (and consequently total lung volume) cannot be measured with spirometry. FROM: West, 1995 with permission.*

A list of the common lung volumes is given in Table 3-1.

*Table 3-1. Common parameters of lung capacity*

| *Lung volume measurement* | *Abbreviation* | *Units* |
|---|---|---|
| *Total Lung Capacity* | *TLC* | *liters* |
| *Inspiratory Capacity* | *IC* | *liters* |
| *Forced Vital Capacity* | *FVC* | *liters* |
| *Functional Residual Capacity* | *FRC* | *liters* |
| *Residual Volume* | *RV* | *liters* |
| *Forced Expiratory Volume in One Second* | $FEV_1$ | *liters* |
| *Lung flow measurement* | | |
| *Peak Expiratory Flow* | *PEF* | *liters/sec* |
| *Maximum flow at 50 or 75% of FVC* | $V_{max}$ *50& 75* | *liters/sec* |
| *Expiratory flow over the mid section of the FVC* | $FEF_{25\text{-}75\%}$ | *liters/sec* |
| *Lung and Airway Resistance* | | |
| *Resistance of the lung* | $R_L$ | $cmH_2O/L/s$ |
| *Resistance of the airways* | $R_{aw}$ | $cmH_2O/L/s$ |
| *Total respiratory resistance* | $R_T$; $R_{rs}$ | $cmH_2O/L/s$ |

In the chapters on the effects of air pollutants you will see that spirometry is used frequently to assess the effects of these pollutants on the lung. In research settings lung function is sometimes measured using more complex (and possibly more accurate) instruments. One such instrument is called a body plethysmograph, sometimes called a "body box" (see Figure 3-2). A plethysmograph operates on the principle that pressure changes across a screen from inside to outside the box can be electronically transformed into changes in the volume of air moving in or out of the lungs. Thus a plethysmograph can be used to measure all the lung volumes. When a subject exhales all the air that she/he can move out of the lungs, a residual volume of air remains. Due to this residual volume the lungs can be expanded easily at the next breath. Without the residual volume, we would have to force our lung open from a collapsed position with each breath. Residual volume can not be measured using spirometry as it is air that is not exhaled from the lungs. Residual volume can be measured by two different techniques. In a plethysmograph, a subject can be asked to pant against a closed mouthpiece. This compresses the air in the lungs and a pressure/volume curve or line can be recorded. Boyles gas law can be applied to the pressure/volume loop and the volume of air in the thorax can be calculated. The measurement is referred to as the thoracic gas volume, that is, the volume of air in the thorax (or lungs). The measurement is often made after expiration and is used to measure residual volume. Another method for measuring residual volume requires use of trace gases and their distribution within the lung (West, 1995).

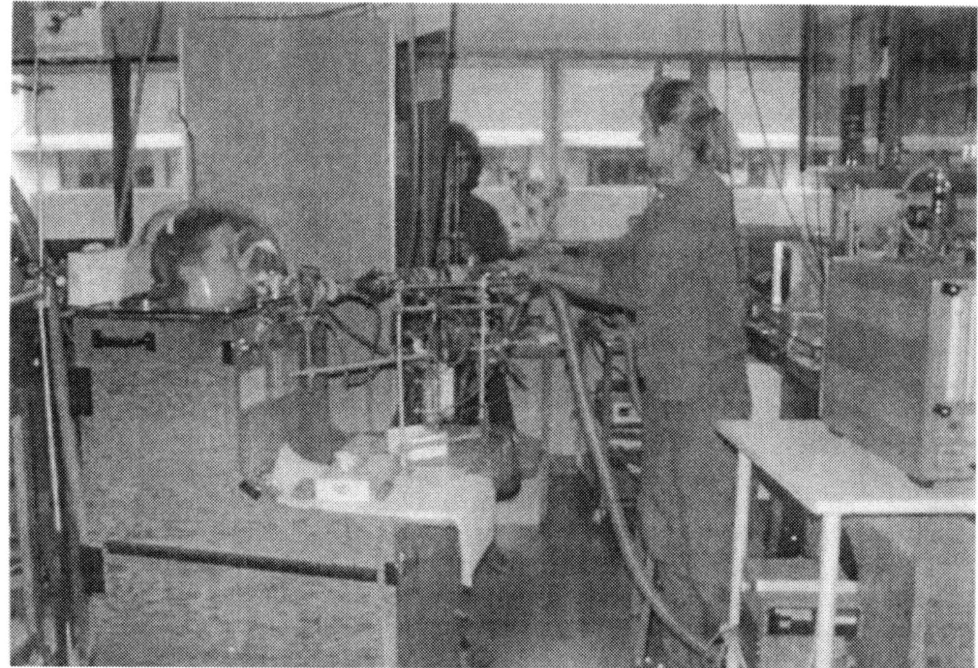

*Figure 3-2. A subject having his lung capacity measured in a pressure-compensated volume-displacement body plethysmograph. (Courtesy Jane Koenig )*

Subject cooperation is important in obtaining valid exhaled lung values. The person who is coaching the subject needs to encourage a robust effort. Subjects are asked to take a deep breath filling their lungs to capacity and then to blow out as hard and as fast and as long as they can. Figure 3-3 shows a young subject preparing to perform spirometry with considerable enthusiasm!

*Figure 3-3. A young subject performing a spirometric maneuver.*

Figures 3-4 and 3-5 show typical nomograms for predicting normal lung function. Note that the nonogram in Figure 3-4 is for normal females 25 years of age or older and the one in Figure 3-5 is for normal males 20 years of age and older.

TO USE NOMOGRAPH: Line up age and height with straight edge and read the predicted values.

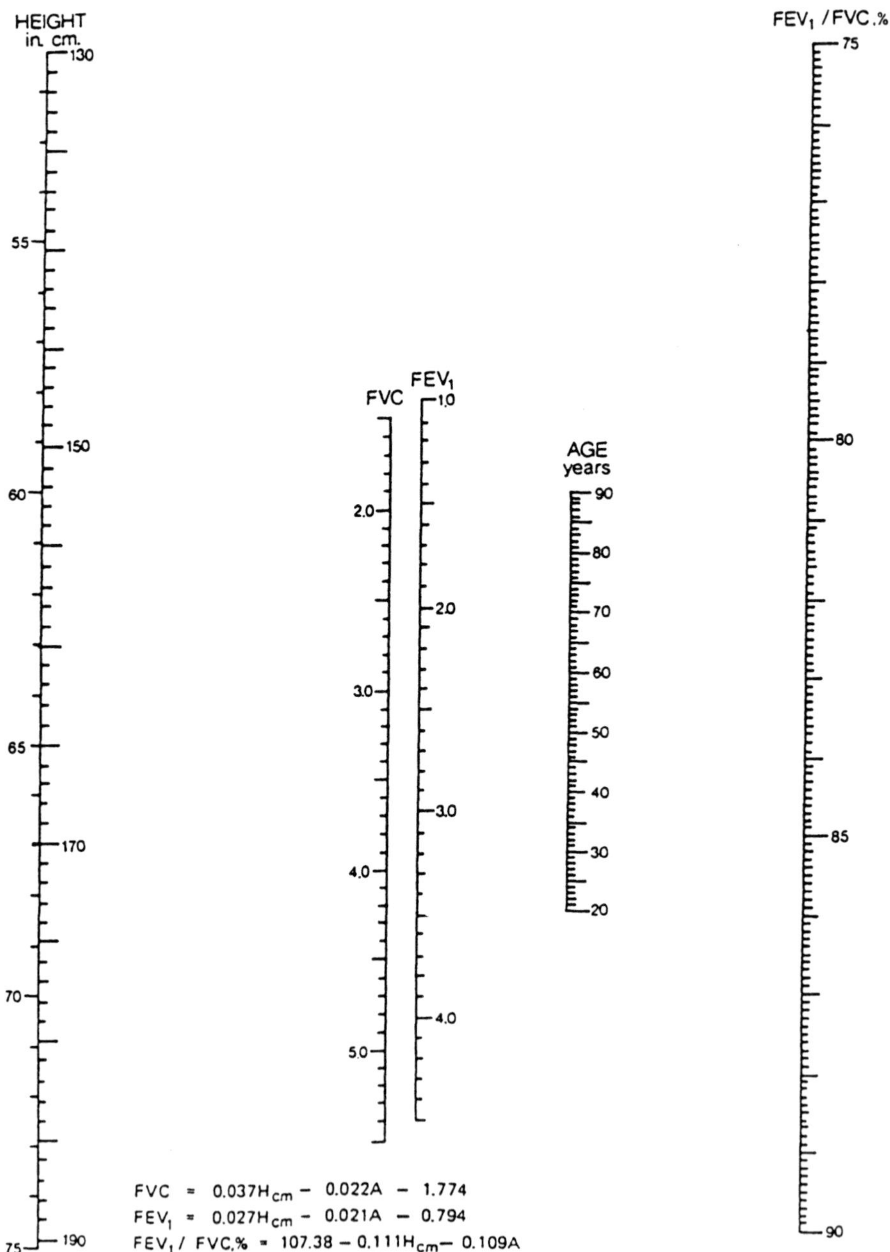


*Figure 3-4. Spirometric nomogram for calculating predicted values of FVC and $FEV_1$ in normal females over the age of 20 years. FROM: NIOSH Spirometry Workbook, 1980.*

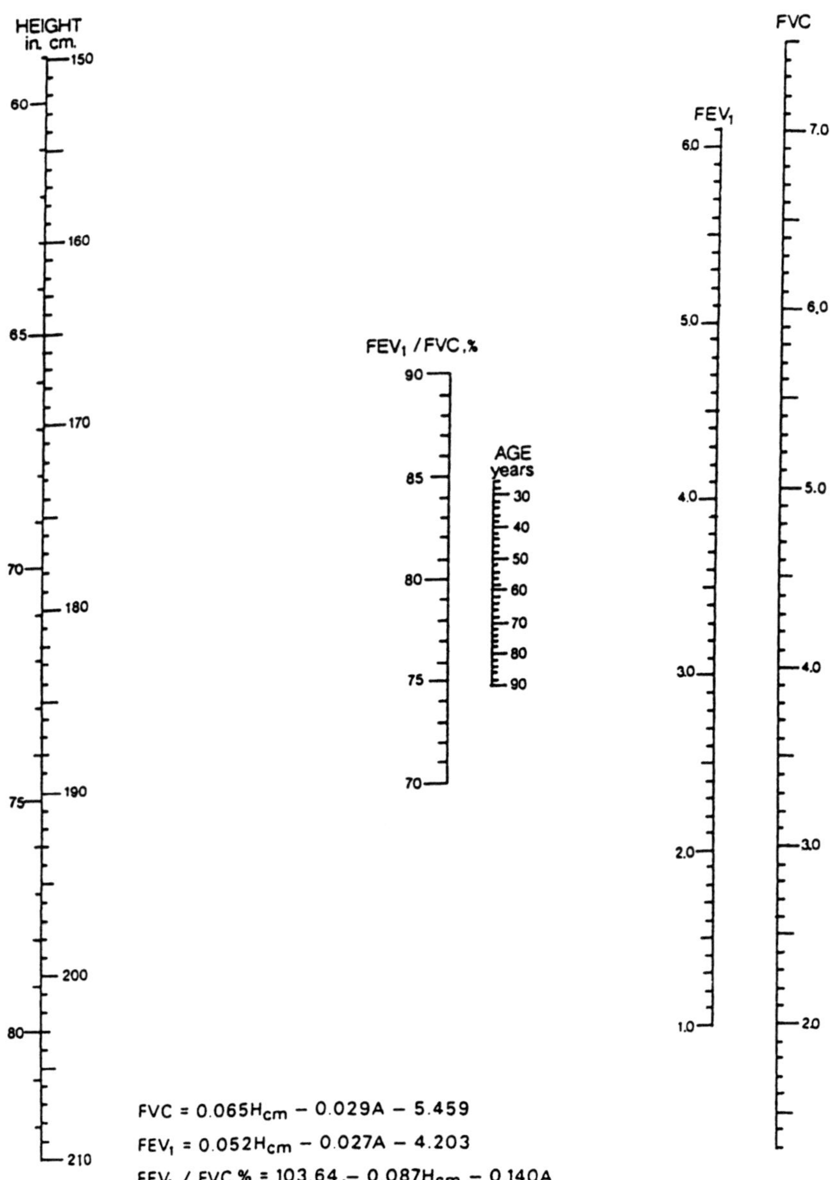


*Figure 3-5. Spirometric nomogram for calculating predicted values of FVC and $FEV_1$ in normal males over the age of 25 years. FROM: NIOSH Spirometry Workbook, 1980*

Exertion increases both breathing rate and inspiratory capacity, thus delivering more oxygen to tissues as respiration needs increase. Some typical ventilatory rates for a variety of activities are shown in Table 3-2.

*Table 3-2. Average minute ventilation rates for a variety of activities*

| *Exercise Level* | *Minute Ventilation (L/min)* | *Activity* |
|---|---|---|
| *Light* | *12-16* | *Level walking at 2 mph;<br>washing clothes* |
| *Light* | *17-23* | *Level walking at 3 mph;<br>scrubbing floors* |
| *Moderate* | *23-30* | *Dancing; pushing wheelbarrow;<br>simple construction;<br>stacking firewood* |
| *Moderate* | *29-38* | *Easy cycling<br>using sledgehammer* |
| *Moderate* | *35-46* | *Climbing stairs;<br>playing tennis;<br>digging with spade* |
| *Heavy* | *42-55* | *Cycling at 13 mph;<br>walking on snow;<br>digging trenches* |
| *Heavy* | *52-57* | *Cross-country skiing;* |
| *Very Heavy* | *62-79* | *rock climbing;* |
| *Very Heavy* | *73-93* | *stair climbing with load;<br>playing squash and handball;<br>chopping with ax* |
| *Very Heavy* | *89-110* | *Level running at 10 mph;<br>competitive cycling* |
| *Severe* | *107-132* | *Competitive long distance running;<br>cross-country skiing* |

It was mentioned earlier that lung volume measurements are very reproducible from time to time in individual subjects. $FEV_1$ is known to vary by 5% or less on repeated measures. However there are situations which can change this. For instance, $FEV_1$ often will be decreased during and after the presence of a lower respiratory viral infection, after exercise (in some people), and after an irritant exposure such a cigarette smoke (or of course, in the context of this book, after exposure to common air pollutants). A comprehensive list of sources of variation in measurement of lung function is given in Table 3-3 (Bates, 1989).

*Table 3-3. Sources of variation in measurements of lung function.*

| *Variation* | | *Source* |
|---|---|---|
| *Within-individual* | *Technical:* | *Within and between instruments.*<br>*Within and between observers in administration and reading of tests*<br>*Curve selection*<br>*Temperature* |
| | *Biologic:* | *Comprehension and/ or co-operation of the subject*<br>*Circadian, weekly and seasonal effects*<br>*Endocrine: other* |
| *Between-individuals* | *All the above sources*<br>+ | |
| | *Subject:* | *Size, sex, age, respiratory muscularity*<br>*Race and other genetic characteristics*<br>*Past and present health*<br>*Habits (e.g., smoking, physical activity)* |
| | *Environmental:* | *Residence (income, ambient pollution, etc.)*<br>*Indoor pollution (smoking, gas stoves, etc.)*<br>*Occupational exposures* |
| *Between-population* | *All the above sources*<br>+ | |
| | | *Selection into or out of the target or study population* |

*FROM Bates DV, 1989 with permission*

Spirometry (or any method of lung function) should not be used for diagnosis of respiratory disease. Lung function measures are a tool in assisting the physician in making a diagnosis of respiratory health. However, there are some relationships that are known to exist between lung function and the two major types of respiratory disease, obstructive lung disease and restrictive lung disease. These are shown in tables 3-4 and 3-5.

*Table 3.4. Relationship of lung volumes to type of ventilatory impairment*

| *Interpretation* | *FVC* | *$FEV_1$* | *$FEV_1/FVC\%$* |
|---|---|---|---|
| *Normal* | *Normal* | *Normal* | *Normal* |
| *Airway obstruction* | *Normal or low* | *Low* | *Low* |
| *Lung restriction* | *Low* | *Low or Normal* | *Normal* |
| *Obstruction & restriction* | *Low* | *Low* | *Low* |

*Adapted and reprinted from Chronic Obstructive Pulmonary Disease, 5th Edition, American Lung Association. 1977.*

*Table 3.5. Spirometric guidelines for assessing degree of ventilatory impairment. Obs = observed; Pred = predicted.*

| | *Obstructive Disease* | *Restrictive Disease* | |
|---|---|---|---|
| | *$FEV_1/FVC\%$* | *$\frac{FEV_1 \text{ Obs.}}{FEV_1 \text{ Pred.}}$ %* | *$\frac{FVC \text{ Obs.}}{FVC \text{ Pred.}}$ %* |
| *Normal* | *≥0.70* | *≥0.80* | *≥0.80* |
| *Mild* | *0.61 - 0.69* | *0.66 - 0.79* | *0.66 - 0.79* |
| *Moderate* | *0.45 - 0.60* | *0.51 - 0.65* | *0.51 - 0.65* |
| *Severe* | *<0.45* | *≤0.50* | *≤0.50* |

*Adapted from Kanner RE. and Morris AH. (editors), Clinical Pulmonary Function Testing, Intermountain Thoracic Society, Salt Lake City, Utah, 1975.*

## NASAL FUNCTION

During quiet inspiration, most people breathe through their nose. As exertion increases, at about 25-30 liters per minute, people switch to mixed nose and mouth inhalation. It has been shown that (without permanent nasal obstruction), inspiration is never totally through the mouth. Breathing through the nose provides the body with several benefits. Air is cleansed of particles and gases and humidified as it travels through the nasal passages. Since dry air is irritating to the lungs of many people, this is an advantage. Also removal of particles and gases serves as a defense mechanism for the lungs. However this defense is not always without adverse consequences for the nose. Inhalation of formaldehyde, a gas that is water soluble and readily taken up by nasal mucous membranes is associated with nasal carcinoma. With respect to air pollution, sulfur dioxide--another water soluble gas--also is taken up readily by the nasal passages. Studies have shown that during quiet breathing up to 99% of the $SO_2$ in the air stream was removed by the nose and not detected by a sensor at the oral pharynx (Speizer and Frank, 1966). $SO_2$ breathed through the nose also has been shown to increase nasal congestion (Koenig et al, 1985). Ozone, a less water soluble gas, is removed less readily. During quiet breathing, approximately 40% of inhaled ozone is removed by nasal surfaces (Gerrity et al, 1988). Deposition and clearance of air pollutants will be discussed in detail later in Chapter 5. Studies from Mexico city are showing that the nasal epithelium in human subjects serves as a sentinel for airborne environmental pollution due to the fact that the nasal passages are often the first site of contact with the environment (Calderon-Garciduenas et al, 1998).

The resistance to air flow through the nose or the work of nasal breathing can be measured by instruments called rhinomanometers. Flow at either the anterior or posterior portion of the nasal passages can be measured. One method is to use a modified scuba mask which has a flow detector mounted on the face of the mask and a pressure port is introduced on the side (see figure 3-6). When a subject wears the mask pressure and flow at each inspiration can be measured. A normal pressure/flow curve from a posterior rhinomanometer instrument is given in Figure 3-7.

*Figure 3-6. A subject demonstrating the use of scuba mask modified to measure the nasal work of breathing.*

*Figure 3-7. Typical pressure-flow curves during measurement of nasal airflow resistance using posterior rhinometry. The tracing is shown on an oscilloscope. Pressure is on the X axis and flow in on the Y axis.*

This summary of physiological measurements of upper and lower respiratory functions should assist the reader in interpreting air pollution-induced changes in these functions. We will see that exposure to each of the common outdoor air pollutants has been associated with adverse effects on lung function or nasal function or both. More information on assessment tools for detecting effects of air pollutants is given in Ch 7.

## REFERENCES

ATS. Standardization of spirometry: 1994 update. Am J Respir Crit Care Med 1995; 152: 1107-1136.

Bates DV. Respiratory function in disease. WB Saunders CO, Philadelphia, 1989. p 132.

Calabrese EJ. Methodological approaches to deriving environmental and occupational health standards. John Wiley and Sons, New York, 1978. P 55/

Calderon-Garciduenas L, Rodriguez-Alcaraz A, Villarreal-Calderon A, et al. Nasal epithelium as a sentinel for airborne environmental pollution. Toxicol Sci 1998; 46: 352-264.

Gerrity TR, Weaver RA, Berntsen J, et al. Extrathoracic and intrathoracic removal of $O_3$ in tidal-breathing humans. J Appl Physiol 1988; 65: 393-400.

Koenig JQ, Morgan MS, Horike M, Pierson WE. The effects of sulfur oxides on nasal and lung function in adolescents with extrinsic asthma. J Allergy Clin Immunol 1985; 76: 813-818.

Speizer FE, Frank R. The uptake and release of $SO_2$ by the human nose. Arch Environ Health 1966; 12: 725-728.

West JB. Respiratory physiology-- the essentials. Fifth Edition. Williams and Wilkins, Baltimore, 1995.

# CHAPTER 4.

# PROPERTIES AND SOURCES OF COMMON AMBIENT AIR POLLUTANTS.

> He was very obliging; and as he handed me into a fly, after superintending the removal of my boxes, I asked him whether there was a great fire anywhere? For the streets were so full of dense brown smoke that scarcely anything was to be seen.
> "Oh dear no, miss," he said. "This is a London particular." I had never heard of such a thing.
> A fog, miss," said the young gentleman.
> "O indeed!" said I.
> Bleak house, Dickens

Air pollution has been around as long as human populations. Most likely one of the first major human causes of air pollution was fire. Along with the invention of fire, we assume, came exposure to various types of smoke. Interestingly, wood smoke is still a primary source of fine particulate matter air pollution in some areas where wood burning is used for home heat or cooking or for clearing of slash.

## SULFUROUS VERSUS PHOTOCHEMICAL AIR POLLUTION

With the advent of the industrial revolution there was a remarkable increase in the burning of coal as a source of energy. As a consequence of this, large amounts of soot and sulfur dioxide ($SO_2$) were emitted into ambient air. This sulfurous type of air pollution dominated in cities until after the second world war. The infamous London killer fog in December 1952 is a prime example of sulfurous air pollution. Around the end of second world war, a different type of air pollution began to be recognized, primarily in Los Angeles. Scientist determined that the cause of the LA air pollution was photochemical atmospheric reactions driven by pollutants emitted from gasoline fueled automobiles. Photochemical air pollution was considerably different than traditional sulfurous air pollution as shown in Table 4-1.

*Table 4-1. Comparison of general characteristics of sulfurous (London) and photochemical (Los Angeles) air pollution.*

| *Characteristics* | *Sulfurous* | *Photochemical* |
|---|---|---|
| *First recognized* | *Centuries ago* | *1944-1948* |
| *Primary pollutants* | *$SO_2$, soot, smoke* | *Ozone, PAN[1], $HNO_3$[2], aldehydes, particulate nitrates and sulfates* |
| *Temperature* | *Cool: < 35°F* | *Hot > 75°F* |
| *Relative humidity* | *High, with fog* | *Low (hot and dry)* |
| *Type of inversion* | *Radiation (ground)* | *Subsidence (overhead)* |
| *Pollutant peaks* | *Early morning* | *Noon-evening* |

[1] *PAN= peroxyacetyl nitrate*
[2] *$HNO_3$ = nitric acid*

*Adapted from: Finlayson-Pitts and Pitts, 1986 with permission*

The contrasts shown in Table 4-1 illustrate many differences in these two types of air pollution conditions. It is easy to recognize that different control strategies are necessary to reduce pollution in both cases. This marked difference noted in 1986 by Finlayson-Pitts and Pitts still exists today. The entire east coast of the US is blanketed in a sulfuric acid haze in summer caused by burning of coal for the generation of electricity. East of the Ohio river, there is rarely 100 kilometers between coal-fired power plants. In contrast the state of California does not have any coal fired power plants in the South Coast Air Basin, the site of Los Angeles. In that region, electricity is generated by burning natural gas or by alternative energy means such as wind power. As a consequence, in California sulfuric acid haze is not a common air pollution problem. On the other hand, in the South Coast Air Basin alone, there are an estimated 6.8 million automobiles, 2.5 million light duty trucks, (producing precursors for ozone formation) and ample ultraviolet light. Southern California is known for the number of sunny days. This combination of automobile emissions and ultra-violet rays from sunlight provide the model conditions for photochemical air pollution.

As a result of major air pollution episodes such as the London killer fog in December, 1952 and a period of air stagnation in Donora, PA associated with excess deaths, the Clean Air Act (CAA) was finally passed in 1970 by the US Congress. The CAA called for the formation of the Environmental Protection Agency (EPA) and authorized EPA to set national ambient air quality standards (NAAQS) for the most common wide spread air pollutants. EPA was also authorized to regulate other air pollutants that were not as wide spread but more often emitted by point sources. These pollutants are described at the end of this chapter. A more detailed discussion of air pollution regulations and standard setting is included in Ch 18 of this book.

## CRITERIA AIR POLLUTANTS

As of 1998, EPA has judged that six common air pollutants are sufficiently wide spread throughout the US that they require regulation on a national basis. These pollutants are carbon monoxide (CO), sulfur dioxide ($SO_2$), ozone ($O_3$), nitrogen dioxide ($NO_2$), particulate matter (PM), and airborne lead (Pb). The common source of all these pollutants is combustion as seen in Table 4.2. They are commonly referred to as "criteria" pollutants.

*Table 4.2 General sources of common wide spread air pollutants*

| | | |
|---|---|---|
| *CO* | = | *Vehicular traffic and other combustion.* |
| *PM* | = | *Wood burning, diesel vehicles and industry.* |
| $O_3$ | = | *Secondary pollutant. Formed from precursors such as auto emissions.* |
| $SO_2$ | = | *Coal fired power plants, smelters, food processing, paper and pulp mills.* |
| *Pb* | = | *Airborne lead mainly from burning leaded gasoline.* |
| $NO_2$ | = | *Combustion (automobiles and industry). Precursor for* $O_3$. |

The "criteria" pollutants are regulated by EPA using NAAQS. They are called "criteria " pollutants because EPA is required to develop a summary of air quality criteria for each of these pollutants every five years in order to up-date the NAAQS if deemed necessary. In July 1997, as a result of such a review, EPA added an additional PM NAAQS for particulate matter (less than or equal to 2.5 micrometers in diameter) and revised the ozone standard both for concentration and averaging time. The reasons for this new PM standard were based on health effects and the fact that fine particles below 2.5 μm are most likely to be products of incomplete combustion. Chapter 10 gives a detailed history of PM regulation in the US. At the same time, EPA changed the NAAQS for ozone from 0.12 ppm averaged over one hour to 0.08 ppm averaged over eight hours. The rationale for this change was that ozone concentrations in the end of the decade of the 1990s are found to peak in the mid-morning and then plateau for the next 8-10 hours. In the 1970s ozone concentrations in cities matched the 1970s pattern of commuter traffic with an ozone peak 1-2 hours after the morning commute, longer concentrations in the early afternoon, and a second ozone peak 1-2 hours after the evening commute. In the 1990s it appears that the freeways are congested most of the day and there is no longer such a clear distinction between the morning and evening commute. Table 4.3 gives the NAAQS regulated by the US EPA NAAQS as of July 1997 along with the target concentration and the averaging times of each pollutant.

*Table 4.3. U.S. National ambient air quality standards* (NAAQS)

| *Pollutant* | *Primary Standard* | *Averaging Time* |
|---|---|---|
| *Sulfur Dioxide ($SO_2$)* | *0.14 ppm* | *24 hours: may be exceeded once per year* |
| | *0.03 ppm* | *1 year* |
| *Particulate matter<10μm ($PM_{10}$)* | *150 μg/m³* | *24 hours* |
| | *50 μg/m³* | *1 year* |
| *Particulate matter<2.5 μm ($PM_{2.5}$)* | *65 μg/m³* | *24 hours* |
| | *15μg/m³* | *1 year* |
| *Ozone ($O_3$)* | *0.8 ppm* | *8 hours* |
| *Nitrogen Dioxide ($NO_2$)* | *0.053 ppm* | *1 year* |
| *Carbon Monoxide (CO)* | *35 ppm* | *1 hour; may be exceeded once per year* |
| | *9 ppm* | *8 hours; same* |
| *Airborne Lead* | *1.5 μg/m³* | *quarterly* |

Four of the criteria pollutants are single gases ($SO_2$, $NO_2$, $O_3$, and CO). All of these except $O_3$ are primary pollutants, that is, they are emitted directly from sources. Ozone in outdoor air, on the other hand, is a secondary pollutant. Ozone formation is covered in more detail in Chapter 11. The two primary sources of precursors for ozone formation are nitrogen oxides and non-methane organic compounds. One of the other criteria pollutants is an element, airborne lead (Pb). Airborne lead was a serious air pollutant in the US until EPA phased out lead as a gasoline additive. Airborne lead is still a serious air pollutant in countries that use leaded gasoline. It is considered a primary air pollutant since it is emitted directly whenever leaded gasoline is burned. Although airborne lead in the US is no longer a major environmental problem, environmental lead in the form of contaminated soil or paint is still a known source of exposure, especially to children. Since lead is not known to be a respiratory irritant, it is not discussed further in this book. The most poorly defined criteria pollutant is particulate matter (PM) air pollution. Particulate matter consists of all the particles and aerosols in the air which are not gases. Particulate matter can be either a primary pollutant (for instance soot or wood smoke particles) or a secondary pollutant such as airborne sulfate and nitrate compounds that are products of atmospheric transformations.

In the US concentrations of gaseous air pollutants are given in parts per million (ppm) and concentrations of PM are given in micrograms per cubic meter ($\mu g/m^3$). However in Europe concentrations of gaseous pollutants are often expressed in $\mu g/m^3$.
Table 4-4 gives conversion factors so the reader can translate between the two metrics.

*Table 4-4. Air quality units conversion table.*

| *Pollutant* | *Multiply ppm by* | *to Obtain* |
|---|---|---|
| *Carbon Monoxide* | *1.145* | *mg/cubic meter* |
| *Ozone* | *1961* | *µg/cubic meter* |
| *Sulfur Dioxide* | *2619* | *µg/cubic meter* |
| *Nitrogen Dioxide* | *1880* | *µg/cubic meter* |

## Carbon Monoxide (CO)

Properties and principles of formation of carbon monoxide. Carbon monoxide was first discovered to be a minor constituent of earth's atmosphere by Migeotte in 1948 (EPA, 1991). By 1993 it was recognized that CO was the most abundant trace gas--other than carbon dioxide--in the atmosphere. Even in so-called pristine regions of the Southern Hemisphere CO concentrations average around 40 ppb (EPA, 1991). Carbon dioxide is a tasteless, colorless diatomic molecule. Its toxic properties are well known. Its major mechanism of action is preferential binding with hemoglobin in blood at the expense of oxygen binding, creating a condition of anoxia. Carbon monoxide is an inevitable consequence of combustion. Thus CO emissions are found in vehicle exhausts, fire place emissions, industrial smoke stack emissions, and in all internal combustion engine emissions. Since the passage of the CAA in 1970 automotive emissions of CO have been reduced more dramatically than any other pollutants. This reduction in shown in Figure 4-1 from the EPA emission trends report (EPA, 1989). It is particularly striking that the CO emissions have decreased so dramatically while vehicle miles traveled in the US has increased at an equally dramatic rate.

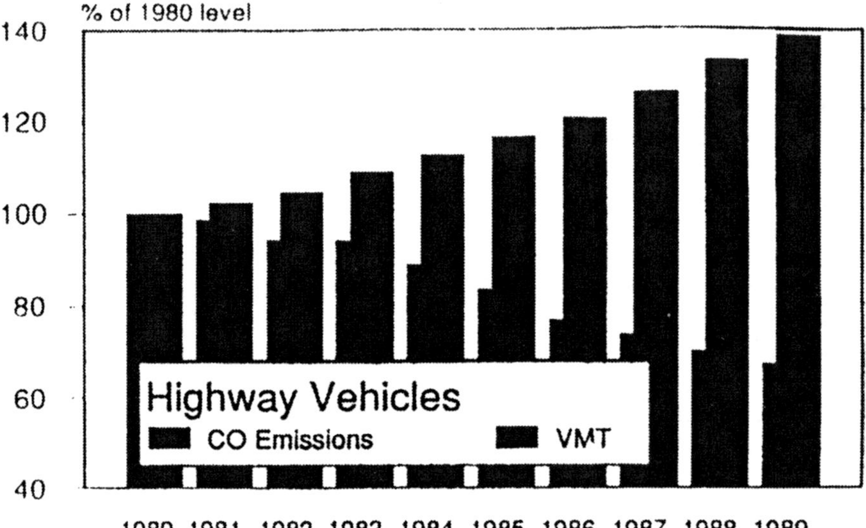


*Figure 4-1 Comparison of trends in total national vehicle miles traveled and national highway CO emissions. FROM: EPA, 1989.*

As can be seen from Figure 4-1, there is a long-term trend toward reduced CO concentrations in ambient air in the US. In fact there has been a 37% decrease in national average outdoor CO levels from 1985 to 1995. These reductions occurred in spite of the 31% increase in vehicle miles traveled. CO emissions have been reduced through better engine technology from automobile manufacturers and through the use of oxygenated fuels that burn more cleanly and thus emit less CO. In spite of this impressive reduction in CO levels, in 1995 there were still 12 million people in the US living in areas that failed to meet the current NAAQS for CO.

## Sulfur Dioxide ($SO_2$)

Properties and principles of formation of sulfur dioxide. Most man-made sulfur oxides ($SO_x$) emissions come from stationary point sources and more than 90% of these discharges are in the form of $SO_2$. $SO_2$ is a colorless gas with a pungent odor. Sulfur oxide emissions can be reduced by controlling the sulfur content of fuels (oil and coal) and using scrubbers to remove the sulfur oxides in the smoke stack. Another technique to reduce point source $SO_x$ emissions has been to build tall smoke stacks which will disperse the $SO_x$ over a wide area. This control technology has created problems of its own by blanketing wide areas with $SO_2$ and sulfate and causing acid deposition throughout much of the eastern coast of the US and southern Canada. Satellite images have shown the entire eastern seaboard covered by a sulfuric acid haze during the summertime. Acid deposition has a deleterious effect on lakes and the fish that live in them and on forests and other types of vegetation. The acids formed from $SO_2$ are also corrosive for buildings and monuments. Ancient statuary in Greece and Italy is being corroded by air pollution containing $SO_2$. $SO_2$ is also emitted from paper and pulp mills, from refineries, and from food processing plants. The highest monitored concentrations of $SO_2$ are recorded in the vicinity of large industrial facilities. The fine sulfate particles that are formed in the atmosphere from chemical reactions between $SO_2$ and water are responsible for degradation of visibility by formation of haze. $SO_2$ is the primary cause of the haze which interferes with vistas in the Western states such as the Grand Canyon in Arizona. $SO_2$ emitted from a coal-fired power plant at Four Corners (where New Mexico, Utah, Arizona, and Colorado are joined) is a major source of that haze. Figure 4-3 shows a schematic model for oxidation of $SO_2$ into sulfuric acid. As can be seen in the figure, the amount of $SO_2$ which stays as sulfuric acid is dependent on neutralizing conditions, primarily the presence of ammonia ($NH_3$) gas. Thus when a $SO_2$ plume passes across agricultural land where there are significant $NH_3$ sources most of the sulfate formed is in a neutralized state (ammonium sulfate or bisulfate). On the other hand, if the $SO_2$ plume travels over an area where there may not be much $NH_3$, the primary transformation product is likely to be sulfuric acid. Neutralization of sulfuric acid is important for both vegetative and human health effects.

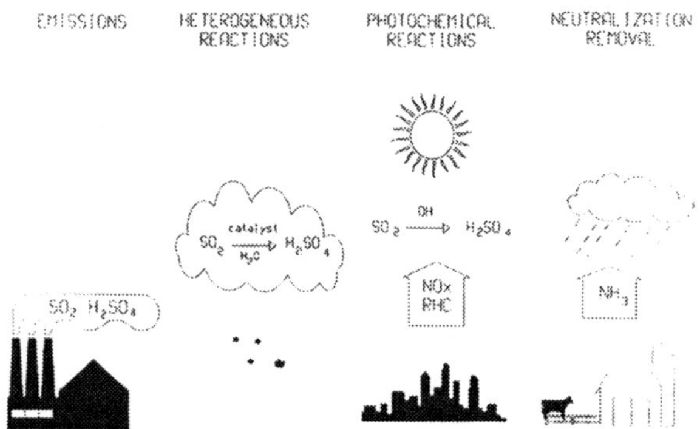


*Figure 4-2. Formation of $H_2SO_4$ from $SO_2$ in the atmosphere. FROM: Dockery and Speizer, 1989 with permission.*

A recent commentary documents that the program of $SO_2$ control instituted by Congress in the early 1990s may be paying off (Kerr, 1998). Congress decided to let power plant operators figure out the cheapest way to control $SO_2$ emissions. Critics of increased $SO_2$ controls projected a price tag of $7 billion increasing to $25 billion by the year 2000.

However experience paints a very different picture. In 1996, $SO_2$ emissions dropped from 10 million tons to 5.4 million tons at a cost of $0.8 billion. This is a true success story in air pollution control. Unfortunately for the case of acid rain, no similar success has been proposed or instituted for the control of $NO_x$ emissions.

There were no counties in the US containing major $SO_2$ point sources that failed to meet the ambient $SO_2$ NAAQS in 1995 (EPA, 1996) based on the 24 hour and annual averages. If there were a short-term averaging time (15 minute or one hour) as part of the NAAQS for $SO_2$ there may have been some exceedances. Washington state, as well as California and Montana, have established short-term standards for $SO_2$ ( Table 4-5).

*Table 4-5. Washington State short-term standards for $SO_2$.*

| | *NAAQS* | *WA state* |
|---|---|---|
| *Annual average* | *0.03* | *0.02* |
| *24 hour average* | *0.14* | *0.10*[a] |
| *3 hour average* | | *0.50*[b] |
| *1 hour average* | | *0.25*[c] |
| *1 hour average* | | *0.40*[a] |

*a = Not to be exceeded more than one per year*
*b = Secondary standard to protect welfare rather than health*
*c = not to be exceeded more than twice in seven consecutive days*

## Nitrogen Dioxide ($NO_2$)

Properties and principles of formation of nitrogen dioxide ($NO_2$). There are eight oxides of nitrogen. Nitrogen dioxide is the nitrogen oxide regulated as a criteria pollutant by EPA. $NO_2$ and nitric oxide (NO) play a major role in the photochemistry of outdoor ozone. Nitrogen dioxide is an odorless, brown gas emitted as a primary pollutant from all combustion processes. $NO_2$ is partially responsible for the yellow/brownish color of polluted air; it is the only ambient gaseous pollutant with significant visible light absorption. Because $NO_2$ is emitted from all combustion sources it is also an indoor air pollution emitted from gas stoves, unvented space heaters, kerosene heaters, wood stoves, and tobacco products as discussed further in Chapter 15 on indoor air problems. Nitrous oxide ($N_2O$) is an important green house gas (see Ch 16). As is the case with $SO_2$, $NO_2$ is transformed in the atmosphere into various acidic products and thus also is a cause of acid deposition. Of the three criteria pollutant gases (CO,$SO_2$ and $NO_2$) only $NO_2$ outdoor concentrations have not decreased since the passage of the CAA in 1970. Figure 4-3 compares the million short tons/year for the primary criteria pollutants (ozone is omitted since it is not a direct emission that can be measured in this fashion). Figure 4-4 shows that $NO_2$ emissions have actually increased since 1970 by 6%.

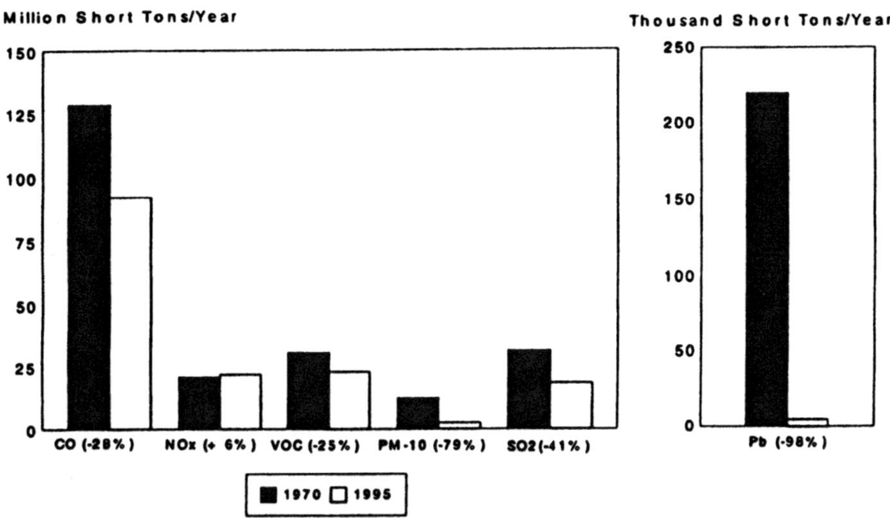


*Figure 4-3. Summary of emissions between 1970 and 1995. FROM: EPA, 1996.*

In spite of the fact that $NO_2$ emissions are not decreasing in the US, there are no areas of the country which exceed the NAAQS for $NO_2$. However that statement needs a bit of interpretation. As shown in Table 4-3, the only NAAQS for $NO_2$ is a concentration averaged annually. That form of averaging time does not allow a careful analysis of short-term (hourly or daily) averages. It would be a mistake to conclude from the fact that there are no exceedances to the NAAQS for $NO_2$ that there are no individuals in the US exposed to unhealthful concentrations of outdoor $NO_2$. There simply are not data that allow us to determine this fact.

## Particulate Matter (PM)

Properties and principles of formation of particulate matter. As mentioned earlier, PM is unique among the criteria pollutants in that it is not a single molecule or element. PM consists of all the dust, haze, and smoke particles in the air. These fine particles are also referred to as aerosols. PM includes both primary particles such as those containing elemental or organic carbon and emitted directly from the tail pipes of gasoline and diesel vehicles and secondary aerosols that are transformed from gases in the atmosphere such as sulfuric acid transformed in the atmosphere from $SO_2$. Particles in the air are found in a wide size range and are typically classified in three sizes; nucleation modes (smallest), condensation mode, and a mechanically generated mode. These size modes and the processes that generate them are depicted in Figures 4-4a and 4-4b. Figure 4-4a shows particle diameter distributions for the fine and coarse modes of PM. The figures are useful for interpretation of health outcome data by PM metric as described in Chapter 10. Figure 4-4b shows the mechanisms generally responsible for each size fraction.

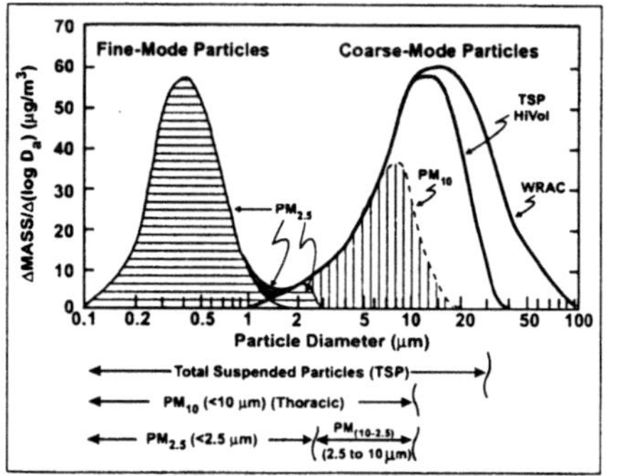


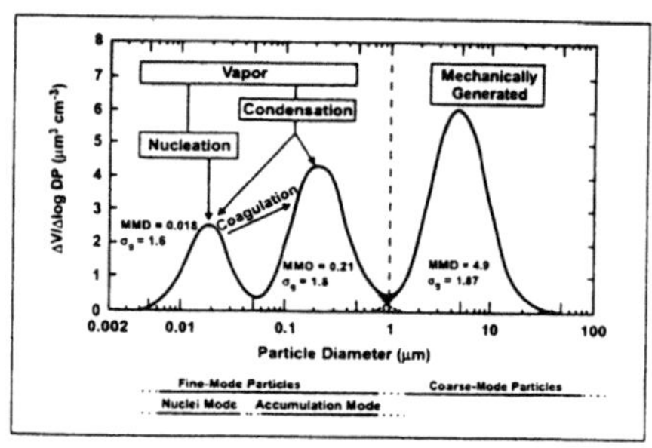


*Figure 4-4. a. An idealized size distribution of ambient PM showing fine and coarse modes and the portions collection in various samples. b. A particulate matter size distribution collected in traffic showing mechanisms for nuclei, fine and coarse modes. FROM: Wilson and Suh, 1998 with permission.*

Nucleation mode particles are usually freshly generated combustion PM having a typical mass mean diameter of 0.02 μm with a geometric standard deviation of 2 μm. Condensation mode particles are mainly combustion derived PM which is composed of condensed materials or smaller particles which have coagulated with a typical mass mean diameter of 0.2 μm with a geometric standard deviation of 2 μm. The mechanically generated particles consist mainly of crustal material (wind blown dust, resuspended dust) and are much larger with a mass mean diameter of 5 μm and again a geometric standard deviation of 2 μm.

Thus there are basically two types of sources of PM: 1) one is natural aerosolization of crustal matter such as during a wind blown dust storm, during agricultural practices, and during excavations for construction which all result in relatively larger size particles closer to 10 μm in diameter. This category also includes resuspended dust from roadways, coal and oil fly ash, sea salt, tire wear debris, and biologic material such as pollen, mold and fungi. 2) The other source of PM is combustion which includes gasoline and diesel fuel vehicle combustion, industry combustion, coal combustion in the process of electricity generation, and burning of vegetative material such as wood burning for residential heating and grass burning for clearing agricultural land. Combustion PM tends to produce significantly smaller diameter particles than crustal PM, in the size range below 2.5 μm-- in fact the average diameter of particles created by residential wood burning is1.0 μm and below. Most of the fine particles are formed from gases by chemical reactions or vaporization. Fine particles are composed of sulfate, nitrate, ammonium and hydrogen ions, organic compounds, and metals. We will see in Chapter 10, that each of these types of fine particles have been implicated as the source of PM-associated health outcomes. The data for each case will be summarized. Natural PM is usually in the size fraction greater than 2.5 μm in diameter. This size fraction is often referred to as the coarse fraction (CF or $PM_{10-2.5}$). Because PM particles are emitted from many different sources, there are large seasonal differences in PM concentrations in many cities and also large differences in $PM_{2.5}$ among cities. In general, in the eastern US a major source of $PM_{2.5}$ is sulfuric acid whereas in the western US a major source of PM is wood or other vegetative burning.

**Ultrafine Particles**

Although particulate matter concentrations in ambient air have been reduced for total suspended particulate (TSP) and $PM_{10}$ (see Figure 4-3), less is known about $PM_{2.5}$ and very little is known about ultrafine particles. Ultrafine (UF) particles are those with diameters less than approximately 0.1 micrometers. The ultrafine particle distribution is represented in Figure 4-4b as the nucleation mode. Human exposure to UF particles is measured by number rather than mass (as $PM_{10}$ and $PM_{2.5}$ are measured). Common number counts for UF particles in US cities are in the range of 50,000 to 100,000 particles per cubic centimeter. A recent article demonstrated the atmospheric organic aerosol formation potential of whole gasoline vapor (Odum et al, 1997). The potential could be accounted for solely in terms of the aromatic fraction of the fuel. This study suggests that gasoline vapors contribute significantly to very small diameter secondary PM. Another recent study measured UF particle emissions from four types of vehicles and found values of 15,000 to 84,000 counts per centimeter cubed (Ristovski et al, 1998).

**Composition of PM**

Scientists try to understand the nature of PM in ambient air by using several different methods. One is source apportionment that is an attempt to define PM by the sources that emit it. Typical sources of PM are mobile sources including cars, buses, and trucks, industrial processes, vegetative burning such as wood burning for residential heat or agricultural burning to clear fields, coal combustion, and incineration. Table 4-6 lists typical sources of both $PM_{2.5}$ and the coarse fraction of PM.

*Table 4-6. Sources of elemental species for $PM_{2.5}$ and $PM_{CF}$ size fractions.*

| *Size Fraction* | *Elements* | *Sources* |
|---|---|---|
| *$PM_{2.5}$* | *Ca, Fe, K, Si, Ti, Al* | *Soil, road dust* |
| | *K* | *Vegetative burning* |
| | *Zn* | *Motor vehicle exhaust, incinerators, residual oil combustion* |
| | *Mn* | *Motor vehicle exhaust* |
| | *Cu* | *Motor vehicle exhaust* |
| | *S* | *Secondary aerosols* |
| *$PM_{CF}$* | *Ca, Fe, K, Si, Mn, Al* | *Soil, road dust* |
| | *Zn* | *Paved road dust, tire wear, blacktop wear* |
| | *Cu* | *Semi-metal brake shoe wear* |

Another method to assist in understanding the chemical and physical nature of PM is reconstruction of the gravimetric mass using the measured species on a filter. This reconstruction provides the major components of PM. Differences in the composition of PM in two regions of the US are shown in Figure 4-5. Note that 47% of the PM is the eastern US is ammonium (from sulfate and nitrate aerosols) whereas the major source of PM in the west is organic carbon (from vegetative burning).

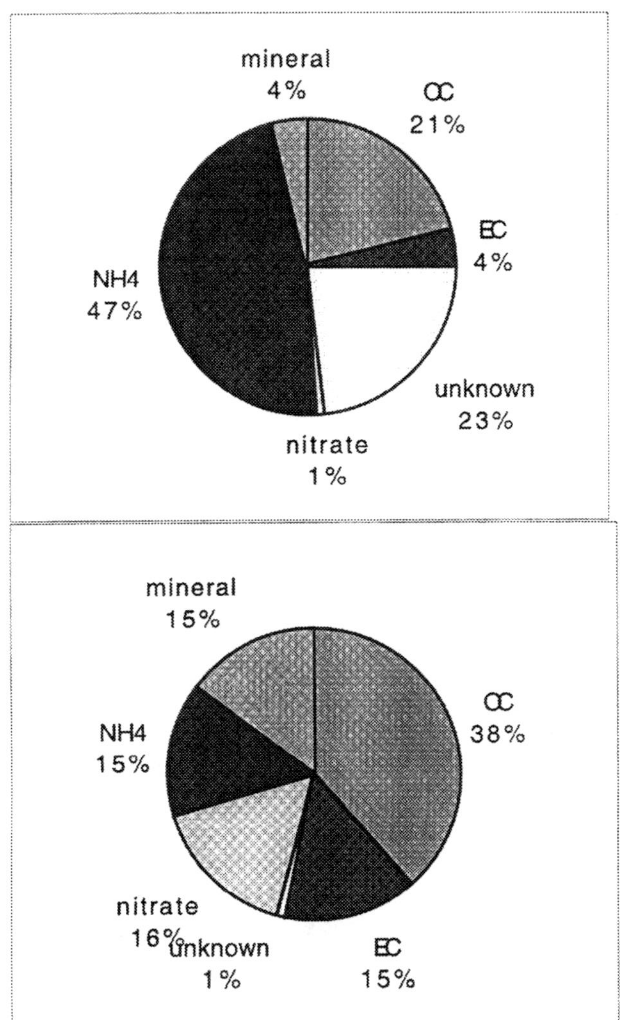


Figure 4-5. Pie charts comparing the composition of PM in the eastern US (on the top) to that in the western US (on the bottom).

**Wood Smoke, A Special Case of PM**

In the western US, the use of wood burning for residential heat is a major contributor to ambient PM. Tables 4-7, 4-8, and 4-9 give the average contribution from wood smoke to PM in three different regions.

*Table 4-7. Chemical mass balance (CMB) results for aggregated wintertime high $PM_{10}$ samples from San Jose, 11/93 to 1/94. (Personal communication, David Fairly)*

| Source | $\mu g/m^3$ | CMB % |
|---|---|---|
| Woodburning/cooking | 36 | 42 |
| Motor vehicle | 15 | 17 |
| Geological | 12 | 14 |
| Marine air | 2 | 3 |
| Sulfate | 2 | 3 |
| Nitrate | 19 | 22 |

In summary, this analysis shows that 42% of the $PM_{10}$ during winter months comes from woodburning and cooking.

*Table 4-8. Chemical mass balance date for particulate matter air pollutants in Seattle, Washington.*

| Source | CMB % |
|---|---|
| Woodburning/cooking | 32 |
| Motor vehicles | 36 |
| Geological | 10 |
| Marine air | 1 |
| Sulfate | 6 |
| Fossil fuel | 15 |

The CMB reports 32% of the PM in Seattle from wood burning (PSAPCA, 1997). However another estimate of source contributions to airborne particulate mass found that in the nighttime during the heating season, 82% of the particle mass was from wood burning, 17 % from mobile sources, and 1% from marine sea salt (Larson et al., 1988).

*Table 4-9. Chemical mass balance data for particulate matter air pollutants in Spokane, Washington. (Norris, 1998).*

| | |
|---|---|
| Organic Carbon | 39% |
| Mineral | 10% |
| Sulfate | 3 % |
| Elemental Carbon | 15% |
| Nitrate | 16% |

Organic carbon (OC) is a marker for vegetative burning which is primarily residential wood burning in the winter months. OC concentrations are elevated in the winter months. Elemental carbon (EC) is a marker for mobile source and stays at a constant value across the four seasons.

**Diesel, Another Special Case of PM**

Another major source of PM emissions into ambient air is diesel vehicles. Diesel trucks and buses do not have to meet the stringent inspection/maintenance standards established for automobiles and light duty trucks. Table 4-10 gives a list of pollutants emitted from combustion of diesel fuel.

*Table 4-10. Abbreviated list of classes of compounds in diesel exhaust*

| | |
|---|---|
| *Particulate Phase* | |
| *Elemental carbon* | |
| *Heterocyclics, hydrocarbons ($C_{14}$-$C_{35}$), and* | |
| *polycyclic aromatic hydrocarbon and derivatives:* | |
| *Acids* | *Ketones* |
| *Alcohols* | *Nitriles* |
| *Aldehydes* | *Quinones* |
| *Anhydrides* | *Sulfonates* |
| *Esters* | *Halogenated and nitrated cpds.* |
| *Inorganic sulfates and nitrates* | |
| *Metals* | |
| *Gas and Vapor Phases* | |
| *Acrolein* | |
| *Ammonia* | |
| *Carbon dioxide* | |
| *Carbon monoxide* | |
| *Benzene* | |
| *1,3-Butadiene* | |
| *Formaldehyde* | |
| *Formic acid* | |
| *Heterocyclics, hydrocarbons ($C_1$-$C_{18}$),* | |
| *and derivatives (as listed above)* | |
| *Hydrogen cyanide* | |
| *Hydrogen sulfide* | |
| *Methane* | |
| *Methanol* | |
| *Nitric and nitrous acids* | |
| *Nitrogen oxides* | |
| *Sulfur dioxide* | |
| *Toluene* | |
| *Water* | |

*FROM: Mauderly, 1992 with permission*

It can be noted that diesel exhaust contains both particulate phase and gas phase pollutants and should be considered a complex mixture of pollutants similar to wood smoke and ETS (Mauderly, 1992).

Much of the exposure to diesel in urban populations comes from transit buses. At project at the University of Washington made recordings with nephelometers (instruments that measure light scattering of fine particles) on both sides of a paved roadway through campus frequented by diesel buses. Figure 4-6 shows the recordings for a 30-minute period. The asterisks show the times that a diesel bus passed the recording site.

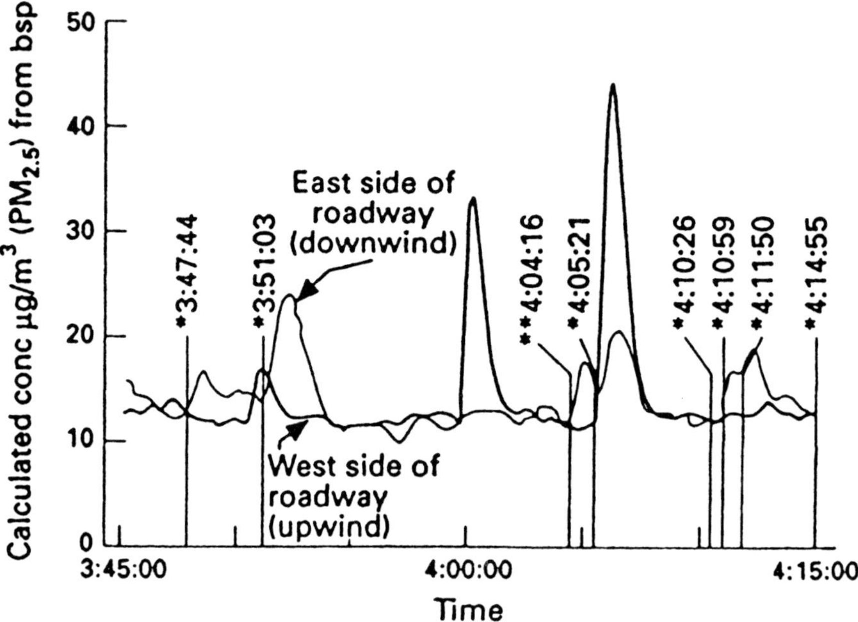


*Figure 4-6. Light scattering recording of fine particles associated with diesel bus emissions. FROM: Balogh et al, 1993 with permission.*

## Ozone

Ozone is a unique air pollutant among the criteria pollutants in that it is the only outdoor pollutant that is solely a secondary pollutant. As can be seen in Table 4-11, ozone is generated from various precursors such as hydrocarbons emitted from automobiles and other vehicles or even from terpines emitted by trees. These precursors react photochemically with NO in the present of ultraviolet light from the sun to form ozone. Ozone concentrations in ambient air generally lag the emissions of precursors by 1-2 hours. Ozone concentrations thus are frequently found several kilometers downwind from the major locations of precursors, the interstate highways and other congested roadways. Another factor that prevents ozone concentrations from building up close to sources of precursors is the highly complex photochemistry involved in ozone production. If sufficient $NO_2$ is present it will serve to scavange the ozone, reducing it as fast as it is produced. Again this explains the location of ozone concentrations downwind from the highways.

*Table 4-11. Formation of ozone, a secondary pollutant, from precursors*

*NonMethane Organic Compounds(MNOC) + $NO_x$ + $hv$ = $O_3$*
*+ other pollutants*

*Sources of NMOC or NMHC:*
*Combustion (primarily mobile sources)*
*Chemical processing*
*Gasoline pumping*
*Oil refining*
*Dry cleaning*
*Terpines from trees*
*Sources of $NO_x$:*
*Primarily combustion, both mobile and stationary*
*Other photochemical oxidants formed by photo-oxidation:*
*Peroxyacetyl nitrate (PAN)*
*Nitrous acid (HONO)*
*Nitric acid ($HNO_3$)*

*$hv$ = ultraviolet light*

## Hazardous Air Pollutants

There also are many air pollutants that are not included in the six criteria pollutants but nevertheless cause, or may cause, severe health effects or ecologic damage. These pollutants are called Hazardous Air Pollutants (HAPs). In the 1990 amendments to the CAA these pollutants were termed "Air Toxics". In almost twenty years, EPA had only set regulations for mercury, beryllium, asbestos, vinyl chlorine, radionuclides, and benzene. The lack of regulation reflects the fact that little information on health effects and outdoor concentrations of HAPs is available. Efforts to assess health effects of HAPs are hindered by the large number of HAPs (>189), multiple exposure pathways, and the lack of long-term monitoring of HAPs in communities. Some of the major HAPs are listed in a toxic release inventory (TRI) operated by EPA; however this list is only for point sources and excludes both mobile and area sources. In the 1995 EPA trends report HAP emissions were estimated to be 39% mobile, 31% area, 16% point sources other than the TRI pollutants, and 14% point sources included in the TRI. So, as with the criteria pollutants, it can be seen that combustion (mobile sources) is a major contributor to HAPs. There are a variety of lists attempting to rank the importance of HAPs from a health effects perspective. Some of these are given here.

*Table 4-12. Major industrial hazardous air pollutants*

| | |
|---|---|
| *Acrylonitrile* | *Freon 13* |
| *Arsenic* | *Hexachloro-cyclopentadiene* |
| *Asbestos* | *Manganese* |
| *Benzene* | *Mercury* |
| *Beryllium* | *Methyl chloroform* |
| *1,3 Butadiene* | *Methylene chloride* |
| *Cadmium* | *Nickel* |
| *Carbon Tetrachloride* | *Perchloroethylene* |
| *Chlorobenzenes* | *Polycyclic organic matter* |
| *Chloroform* | *Radionuclides* |
| *Chromium* | *Toluene* |
| *Coke Oven Emissions* | *Trichloroethylene* |
| *Dioxin* | *Vinylidene* |
| *Epichlorohydrin* | *Vinyl chloride* |
| *Ethylene dichloride* | *Metals* |
| *Ethylene oxide* | |

---

*Table 4-13. List of priority HAPs (EPA) judged to be major and increasing threats to human health and the environment.*

*1,3-butadiene*
*acetaldehyde*
*acrolein*
*benzene*
*beryllium*
*carbon tetrachloride*
*chlorine*
*chloroform*
*chloroprene*
*chromium VI*
*dioxin*
*formaldehyde*
*furans*
*mercury*
*methylene chloride*
*styrene*
*tetrachloroethylene*
*toluene*
*vinyl chloride*
*xylene*

EPA has published a list of the top 20 sources of HAPs (see Table 4-14). The source that contributes the most HAPs is on-road vehicles. Residential wood burning is second in amount of emissions. Residential wood burning also is a major source of $PM_{2.5}$. This fact has been recognized by air agencies in Western states that call bans on wood stove use when air stagnations occur and air pollution levels rise. Only nine of the top 20 sources of HAPs are industrial.

*Table 4-14. Top 20 sources of toxic emissions for 37 toxic pollutants, 1990: Total annual emissions in tons/year.*

| *Rank* | *Category Description* | *Tons/year* |
|---|---|---|
| *1* | *On-Road motor vehicles* | *1.52E+06* |
| *2* | *Residential wood combustion* | *5.25E+05* |
| *3* | *Glycol dehydrators* | *2.45E+05* |
| *4* | *Consumer and commercial product solvent use* | *2.22E+05* |
| *5* | *Non-road mobile vehicles* | *2.09E+05* |
| *6* | *Forest files* | *1.91E+05* |
| *7* | *Prescribed burning* | *1.31E+05* |
| *8* | *Industrial wood waste combustion* | *9.93E+04* |
| *9* | *Dry cleaning* | *8.98E+04* |
| *10* | *Halogenated solvent cleaning* | *5.77E+04* |
| *11* | *Utility coal combustion* | *3.96E+04* |
| *12* | *Gasoline distribution; stage II* | *2.27E+04* |
| *13* | *Primary aluminum production* | *1.80E+04* |
| *14* | *Industrial coal combustion* | *1.69E+04* |
| *15* | *Manufacture of motor vehicles and car bodies* | *1.51E+04* |
| *16* | *Gasoline distribution; stage I* | *1.37E+04* |
| *17* | *Plastics foam products* | *1.36E+04* |
| *18* | *Commercial printing, gravure* | *1.27E+04* |
| *19* | *Pulp mills* | *1.21E+04* |
| *20* | *Structure fires* | *1.18E+04* |

*FROM: EPA. National air quality and emissions trends report, 1995.*

## CONCLUSIONS

This intent of this chapter has been to introduce the reader to common sources of both criteria and hazardous air pollutants. It is suggested that readers review the pertinent sections in Ch 4 as they read about the associated health effects in later chapters.

# REFERENCES

Balogh M, Larson T, Mannering F. Analysis of fine particulate matter near urban roadways. Transportation Res Record 1993; 1416: 25-32.

Dockery DW, Speizer FE. Epidemiological evidence for aggravation and promotion of COPD by acid air pollution. Lung Biol Health Dis 1989; 43: 201-225.

EPA. Air quality criteria for particulate matter and sulfur oxides. EPA/600/8-86/020F, December, 1986.

EPA. Air quality criteria for oxides of nitrogen. EPA/600/8-91/049aA, August 1991.

EPA. Air quality criteria for carbon monoxide. EPA/600/8-90/045F. December, 1991.

EPA. National air quality and emissions trends report, 1989. EPA-450/4-91-003. February 1991.

Finlayson-Pitts BJ, Pitts JN Jr. Atmospheric Chemistry. Fundamental and experimental techniques. Wiley-Interscience, New York, 1986.

Kerr RA. Acid rain control: success on the cheap. Science 1998; 282: 1024-1027.

Larson TV, Kalman D, Wang S-Z, Nothstein H. Urban air toxics mitigation study: Phase I. Report prepared for the Puget Sound Air Pollution Control Agency, Seattle, WA 1989.

Mauderly JL. Diesel exhaust. IN Lippmann M (ed). Environmental toxicants: Human exposures and their health effects. Van Nostrand Reinhold, New York, 1992.

Norris G. Air pollution and the exacerbation of asthma in an arid, Western US city. PhD Thesis, Department of Civil and Environmental Engineering, University of Washington, June 1998.

Odum JR, Jungkamp TPW, Griffin RJ, et al. The atmospheric aerosol-forming potential of whole gasoline vapor. Science 1997; 276; 96-99.

Puget Sound Air Pollution Control Agency. Puget Sound $PM_{10}$ emission inventory: Attachment 2, June, 1997.

Ristovski ZD, Morawska L, Bofinger ND, Hitchins J. Submicometer and supermicrometer particulate emission for spark ignition vehicles. Environ Sci Technol 1998; 32: 3845-3852.

Wilson WE, Suh HH. Fine particles and coarse particles: Concentration relationships relevant to epidemiologic studies. J Air Waste Manage Assoc 1997; 47: 1238-1249.

# CHAPTER 5.

# EXPOSURE ASSESSMENT

Human exposure assessment is a discipline that develops measurement tools and conducts measurements of the contact of an individual or population with contaminants released into the environment. In the realm of air pollution health effects, exposure assessment refers to measurements of air pollutants in situations where humans may be exposed through inhalation. There are three primary types of exposure measurements relied on for assessing an individual's exposure to an air pollutant:

Outdoor monitoring
Indoor monitoring
Personal monitoring

Outdoor air monitoring of criteria pollutants asa requirement for communities is largely conducted by the US EPA. Investigators conducting research often supplement the fixed site monitoring networks managed by air agencies. There is no requirement for indoor air monitoring outside of the workplace or in unusual situations such as asbestos monitoring in school buildings. Likewise, personal monitoring outside of the workplace is not conducted in any systematic manner. Both indoor monitoring and personal monitoring for air pollutants are conducted solely for ad hoc or research purposes.

Exposure assessment involves determining the sources of pollutants, their transport and fate, exposure of populations (or whatever target the investigator decides), the dose of the pollutant to target tissues, and ultimately the health effects. Recall that exposure is defined as the concentration of the pollutant in a given environment and dose is defined as the concentration within the body at the target tissue. Sources of pollutants and their transport and fate were covered in Ch 4. The health effects of individual pollutants are dealt with in later chapters.

It is important to spend some time understanding the strengths and weaknesses of present day exposure assessments for individuals and populations, since epidemiology of the health effects of air pollution is almost entirely dependent on the quality of the exposure assessment in a given study. The National Academy of Sciences reviewed human exposure assessment for airborne pollutants in 1991 (NRC, 1991) and that source is recommended for those readers interested in more depth on this issue.

As will be seen later in this book when we review studies documenting adverse health effects of air pollutants, exposure measurements used in these studies ranges from the presence of one fixed site air pollutant monitor for an entire population to actual indoor air measurements for subjects to personal samplers worn for 24 hours or more. Measurement of the actual dose at target tissues is rare and will be discussed somewhat later in this chapter when we discuss biomarkers. Obviously the accuracy of the exposure estimate is better if personal sampling data or at least indoor air monitoring data are available than if the investigator relies on the values from the fixed site monitor. Concentrations of air pollutants are measured in milligrams or micrograms/ meter cubed ($mg/m^3$ or $\mu g/m^3$) for particles and parts per million or parts per billion (ppm or ppb) for gases.

As well as direct methods described above, exposure assessment methodology uses indirect techniques including questionnaires asking subjects to recall (retrospective) or record their exposure and activities on a daily basis (prospective). Biological markers of exposure and modeling of pollutant concentrations are also used to estimate exposure. The pollutant of interest and its physical, chemical, and biological characteristics, as well as the hypothesis being tested, greatly determine the type of exposure assessment available. This can be seen readily in the field of air pollution epidemiology. The criteria pollutants can be as easy to measure as carbon monoxide (a single gaseous molecule which is not very reactive) or as difficult as particulate matter where measurement of the mass on a filter for a 24 period is the norm. The measurement of mass concentration gives the investigator no information about the chemical constituents in that mass. A sample that was 100% wood smoke and another that was 100% diesel exhaust would be indistinguishable by mass measurement alone. If the investigator wants to know about the chemical and physical properties of PM, various chemical assays must be performed. These are usually expensive both in terms of time and funds. The chemical and physical properties of the pollutant (only PM for the criteria pollutants) also determine the method used for personal samplers. Personal samplers are available both as passive samplers which work for diffusive gases to active samplers which are required for particles and aerosols. Active samplers require a pump and thus become heavier and nosier than simple passive samplers. Passive samplers are well suited for longer term integrated measures of the pollutant. However if short-term peak values are driving the health endpoint, they may be of little use. The field of personal samplers is one where advances can be expected each year. An example of data collected in a study of personal exposures to $NO_2$ is given in Table 5-1. The data in this table were collected using Yanigisawa personal $NO_2$ badges (Yanagisawa and Hishimura, 1982). The children in this study ranged in age from three to sixteen years of age and were divided into two main groups: a control group of children residing in homes that use electric appliances as the primary source of cooking, and a study group of children residing in homes that use gas cooking appliances. The mean concentrations for the exposed and unexposed groups were 12.0 and 5.6 ppb $NO_2$ respectively.

*Table 5-1. Summary of passive sampling results for $NO_2$ in eight children*

| *Exposed Group* | | *Gas range* | *Gas Oven* | *Unexposed* | |
|---|---|---|---|---|---|
| | *$NO_2$* | *minutes* | *minutes* | | *$NO_2$* |
| *Subject #* | *(ppb)* | | | *Subject #* | *(ppb)* |
| *12,14* | *20.9* | *69* | *94* | *8A* | *5.2* |
| *13* | *5.7* | *0* | *0* | *8B* | *8.8* |
| *2,9* | *9.5* | *11* | *0* | *1* | *3.5* |
| *3,6* | *11.9* | *720* | *95* | *7* | *4.9* |
| *Mean* | *12.0* | | | *Mean* | *5.6* |

The nitrogen dioxide pilot study using passive badge sampling clearly showed a relationship between exposure levels in homes that use gas appliances and those that use electric. The results revealed children in the study group (exposed) as having exposures levels twice as high as those children in the control group (unexposed). (Norris, Murphy and Koenig, unpublished data).

Accurate exposure assessment requires that all sources and routes of exposure are measured. This has led to a idealized state of knowledge referred to as the Total Exposure Assessment Method. The many variables entering into calculation of total exposure to an air pollutant are shown in the diagram in Figure 5-1. Even a 24-hour active sampler and chemical analysis are not adequate to achieve total exposure. One also needs to know whether any exposure could have occurred by ingestion or skin penetration, though this is unlikely with the common air pollutants. In addition, one needs to know the activity level of the individual being studied since ventilation rate is one of the determinants of dose.

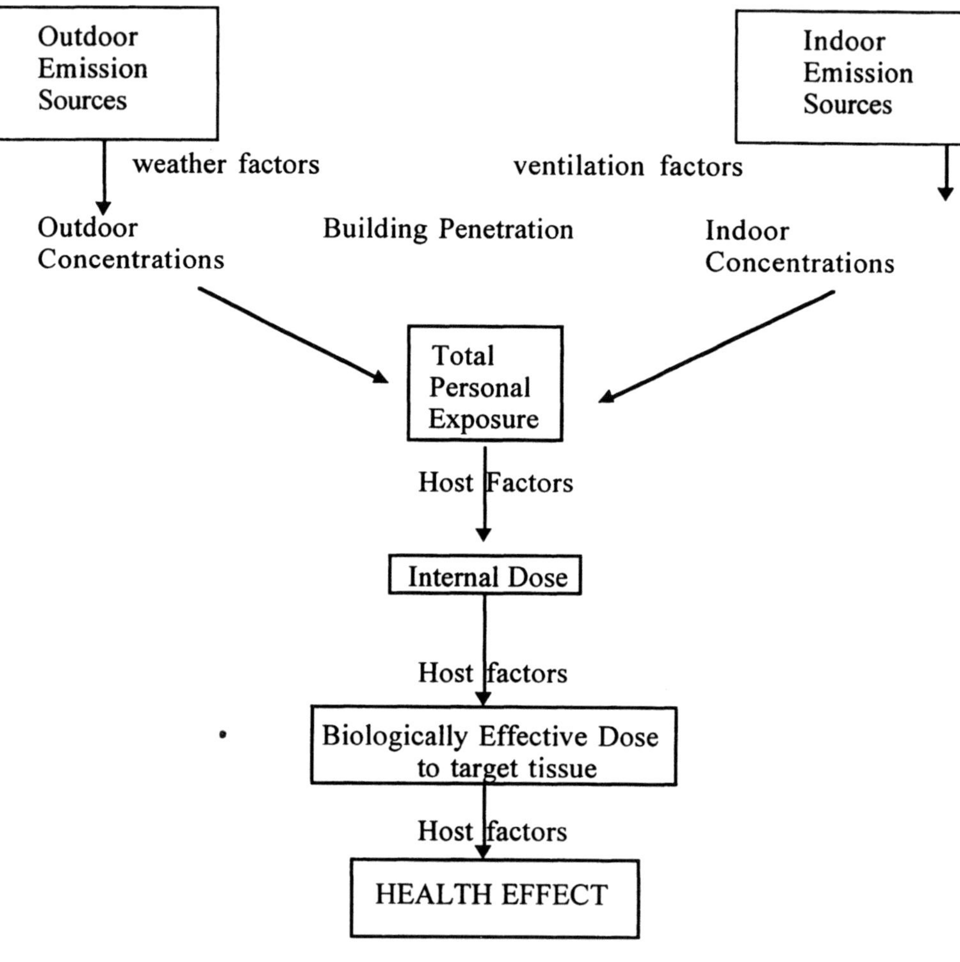


*Figure 5-1. A framework for exposure assessment.*

Even a well carried out exposure assessment including personal samplers does not allow the investigator to specify the actual dose of the air pollutant which reaches the target tissue. In fact, the internal dose may not be equal to the biologically effective dose. The difference in the exposure concentration and the concentration of biologically effective dose will depend on uptake of the gas or deposition of the particle, distribution in the airways (upper and lower), and metabolism at the very least. However, a basic estimation of dose, if one knows the exposure, is as follows:

Dose = concentration X duration X ventilatory rate

Therefore, when a subject is exposed a given concentration of ozone in a controlled laboratory setting, we can calculate dose if we know the durationof exposure, the ventilatory rate and the fraction of the agent retained in the lung. For instance, if a subject is exposed to 0.12 ozone for 30 minutes while exercising and breathing at a rate of at 24 l/minute, an estimation of dose would be

Dose = 0.12 ppm X 30 min X 24 l/min = 86.4 ppm liters ozone

Obviously this is only an estimate. However for ozone, under some conditions, it may be fairly close to the actual dose since studies have shown that approximately 90% of inhaled ozone in retained within the respiratory tract.

Time activity charts are used to help estimate the total exposure. Table 5-2 shows a timer activity chart and instructions on its use, that was developed for use with subjects with asthma. This table obviously is truncated to save room. If actually used, the table would have as many rows as the investigator required. For example, one row for 6-8 am, one row for commuting, one row for 9-12 am, one row for activity during lunch break, one row for afternoon commute, two rows for 6pm until bedtime, and one row for hours asleep.

*Table 5-2. Time activity and daily diary.*
*ID#:* *Date:* *Meter #*
*Morning Peak Flow:* ____________ ____________

| | *Place an X in the Appropriate Category* | | | | | |
|---|---|---|---|---|---|---|
| *Military Start Time* | *Home (Indoors)* | *Out doors* | *School or Indoor Camp* | *Travel* | *Other (comment)* | *Asthma Medication Taken* |
| | | | | | | |
| | | | | | | |
| | | | | | | |
| | | | | | | |
| | | | | | | |

*Evening Peak Flow:* ____________ ______

*Summary (check all that apply):*

| | |
|---|---|
| _______*Trouble breathing* | _______*Fever* |
| _______*Visited doctor/nurse* | |
| _______*Runny/stuffy nose* | _______*Burning or red eyes* |
| _______*Cough* | _______*Upset stomach* |
| _______*Visited ER* | |
| _______*Wheezing* | _______*No symptoms* |
| _______*Hospitalized* | |

*Instructions for Using the 24-hour Time Card*
*(Record the date on your form; use a new card for each date)*
*1. Take and record the three morning peak flow measurements before breakfast.*
*2. Each time your location changes (leaving home, arriving at school, etc.), please record the day and military (24-hour) time on the time card. Place an X in the column describing the location/activity that you are now beginning. Record all locations from waking in the morning through to the following morning on this card.*
*3. Time at SCHOOL includes time doing your usual school activities while inside and outside your school building or area.*
*4. Time OUTDOORS includes all time spent outdoors except for routine recess times at school.*
*5. OTHER includes any other activities away from your home and school that are not conducted outside. A brief commentary on your location will be helpful to the study.*
*6. HOME includes time spent indoors in your home or garage. Time spent in the yard of your home is OUTDOORS.*
*7. TRAVEL includes time spent commuting to school and other places.*
*8. Please mark an X under ASTHMA MEDICATION TAKEN for any time periods where asthma medication is used.*
*9. Take and record the three evening peak flow measurements at the time of teeth brushing before going to bed.*
*10. Check all symptoms and events that you experienced from the time you wake in the morning until rising the following morning.*
*11. If you have any problems or questions, contact the investigator.*

---

An instructive class exercise would be for students to record their own data in the time-activity diary. This would give them an idea of the commitment and accuracy associated with this measurement tool.

Another method of exposure assessment is the use of a recall questionnaire. However, problems are often associated with recall data, as studies have shown that recall is not very accurate and may be prone to biases. Examples of questions that might be included on such a questionnaire are:

Were you exposed to environmental tobacco smoke (ETS) during the past two weeks?

How many hours per day do you spend commuting?

Have you been exposed to diesel exhaust during the past two weeks?

How many hours did you spend in vigorous activities in the past two weeks?

Indeed it is difficult to answer such questions accurately. A typical exposure profile (for $NO_2$ in this example) derived from a time activity analysis is shown in Figure 5-2. It is apparent that meal preparation and commute are responsible for $NO_2$ exposure in this situation for a subject using a gas cooking stove. As you will see, in the discussion of $NO_2$ health effects in Ch 12, gas cooking stoves emit considerable concentrations of nitrogen oxides.

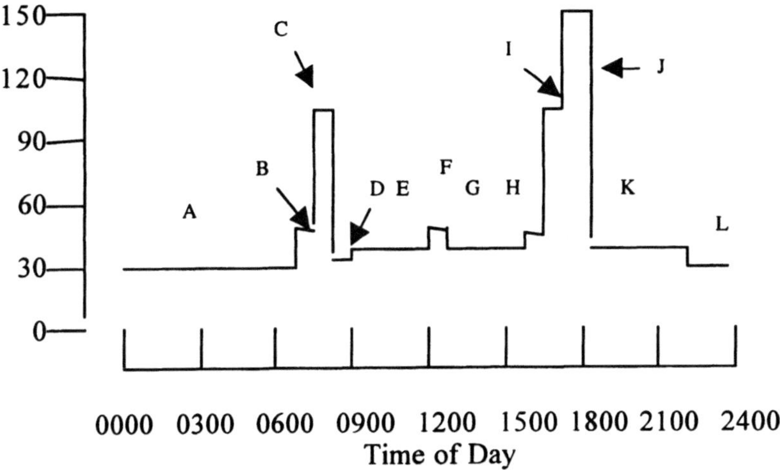


*Figure 5-2. An exposure profile for $NO_2$ showing the effect of different activities. The code is: A = Home, asleep; B = Meal preparation; C = Commute; D = Office, early morning; E = Office, later morning; F = Lunch, restaurant; G = Office, afternoon; H = Office, late afternoon; I = Commute; J = Meal preparation; K = Home, evening; L = Home, asleep. Adapted from Sexton and Ryan, 1988.*

Determination of an exposure assessment requires estimates of the size and susceptibility of the population exposed as well as the estimates of exposure concentration of the pollutants described above. No matter how toxic a given air pollutant may be, there is no health effect if there is no one breathing the air at that time--that is, if there is no exposure. It can easily be seen from Table 5-3 that a large number of individuals in the US are actually exposed to the criteria air pollutants at concentrations beyond the standards. Since this estimate was published the NAAQS standards for both ozone and PM have been tightened, thus increasing the numbers of individuals at risk.

*Table 5-3. Estimated populations at risk residing in communities that have not attained one or more National Ambient Air Quality Standard - United States, 1991*

| Population subgroup at risk | Pollutant | At-risk population living in nonattainment areas No. | (%) |
|---|---|---|---|
| Preadolescent children (aged ≤ 13 yrs) | PM-10, $SO_2$, $O_3$, $NO_2$ | 31,528,939 | (63) |
| Elderly (aged ≥ 65 yrs) | PM-10, $SO_2$, $O_3$ | 18,846,666 | (60) |
| Persons with pediatric asthma | PM-10, $SO_2$, $O_3$, $NO_2$ | 2,285,061 | (61) |
| Adults (aged ≥ 18 yrs) with asthma | PM-10, $SO_2$, $O_3$, $NO_2$ | 4,279,413 | (65) |
| Persons with chronic obstructive pulmonary disease | PM-10, $SO_2$, $O_3$, $NO_2$ | 8,831,970 | (64) |
| Persons with coronary heart disease | CO | 3,493,847 | (33) |
| Pregnant women | CO, Pb | 1,602,045 | (38) |
| Children aged ≤ 5 yrs | Pb | 74,312 | (3) |

*MMWR, 1993.*

## BIOMARKERS

Due to the great expenditure of time, money, and subject commitment in obtaining precise personal samples of air pollutant concentrations, there has been an attempt to develop biomarkers of exposure that could document the amount of a given pollutant actually inhaled. There are three types of biomarkers: exposure, effects, and susceptibility. We will discuss biomarkers of effects and susceptibility in other chapters. Table 5-4 gives a list of ideal characteristics for a biomarker of exposure

*Table 5-4. Advantages of Biomarkers of exposure*

*Improve exposure estimate*
*Incorporate both environmental and personal exposures*
*Can actually estimate DOSE rather than just EXPOSURE*
*Can estimate distribution, metabolism, clearance rates, and biological accumulations*
*Can measure persistence -- retrospective exposure assessment is possible*
*If the biomarker is stable, it can be collected in the field and analyzed later*

Biological markers offer the potential to greatly improve both the quality and efficiency of exposure estimates for population studies and studies of pulmonary effects in relation to air pollution episodes. The most useful biomarker would be specific to one air pollutant. Biological markers of exposure incorporate both environmental exposure factors (air pollutant levels) and personal exposure factors (ventilation rate, personal microenvironmental and/or behavior-mediated air levels, and personal toxicokinetic factors affecting uptake, distribution, metabolism, clearance rates, and resulting biological accumulations). Because biological levels of marker compounds persist for hours to days following inhaled exposures, outcome-initiated sampling and retrospective exposure assessment are possible. This permits study of prior exposures in sensitive populations under conditions when triggering events (onset of symptoms, doctor visits, hospital admissions) have occurred, without the need to prospectively characterize exposures in a large study group in order to capture a limited number of triggering events.

Sources of biomarkers of effects can be both nasal and bronchoalveolar lavage fluids, blood, urine, hair, sputum, teeth, and bone. Obviously some of these tissues would give estimates of more acute exposure (urine, blood, lavage) while other would give more chronic or body burden estimates (bone and teeth). Bone biopsies are being used as biomarkers of exposure to lead. Teeth (especially baby teeth) have been used as biomarkers for lead exposure in children. Hair is a common biomarker for chronic exposure to arsenic. Urinary levels of arsenic can be used for estimates of more acute exposure. A more detailed description of biomarkers as assessment tools is given in Ch 7.

The prototypic biomarker of exposure in the air pollution field is carboxyhemoglobin (COHb) which gives an excellent estimate of recent exposure to carbon monoxide. Carbon monoxide binds preferentially with hemoglobin thus reducing the oxygen carrying content of blood. The net result is a reduction of $O_2$ availability and hypoxia in affected tissues. A simple blood assay can determine the percentage of COHb. Ordinarily the normal COHb percentages in human blood are less than 1%. Cigarette smokers have levels of COHb of 5%. Exposure to 100 ppm CO in a controlled exposure for one hour during intermittent exercise leads to COHb levels of approximately 4%. CO exposures and health effects are described in Ch 13.

Dave Kaman and associates at the University of Washington have developed a biomarker of exposure to wood smoke (Zhu, 1998). The objective is to use the newly-devised biological marker of wood smoke exposure to augment personal and area air monitors, in order to assess doses received under conditions of known environmental pollution. Urine samples were collected from residents of Texas during the summer of 1998 when wood smoke from Mexico was transported into the US. It is hoped that these samples will help understand the usefulness of the biomarker. This will serve both to improve exposure estimates by including personal factors that modify dose, and will help to further characterize and validate the biological marker for use in future studies.

The compounds used as biomarkers of wood smoke exposure are a group of 11 methoxylated phenols and related oxygenated aromatics. Among the several components of wood smoke that have been considered as tracers for use in atmospheric source apportionment studies, only alkyl methoxylated phenols have the advantage of being present in wood smoke at high concentrations and having no common atmospheric sources other than wood burning. Two groups of these lignin pyrolysate compounds in particular are noteworthy: guaiacols and syringols (Hawthorne et al, 1988). Guaiacols and syringols together comprise about 10-28% of the organic carbon in wood smoke particulate matter based on source samples. This accounts for 5-10% of the fine particulate organic mass in ambient 12-hour samples collected under conditions where wood smoke impacts are known to be relatively high. The type of wood used for burning is predictive of both the total contribution of these compounds to particulate organic matter and of the relative contribution of syringols. These compounds are reasonably well conserved in the atmosphere and correlate well with radiocarbon assays in differentiating wood combustion from fossil fuel sources of atmospheric particulate carbon. Ambient air concentrations of individual or combined guaiacols and syringols have not been widely reported. Hawthorne et al (1988) measured winter time ambient levels in ambient air in a variety of locations in Minneapolis (mainly hardwood fuel) and Salt Lake City (mainly softwood fuel). Three types of urban areas were selected: residential neighborhoods with low traffic; residential with high traffic; and residential-retail areas. Single compounds had concentrations of approximately 500 $ng/m^3$ for a 12 hour period.

Another family of biomarkers of exposure is DNA adducts. DNA adducts are used as biomarkers of internal dose of certain carcinogens such as polycyclic aromatic hydrocarbons (PAHs). Although the focus of this text is noncancer endpoints of air pollution exposure, it is important for the reader to be aware of this method for determining personal exposure. A study by Costa and associaties (1998) used DNA adducts to estimate the dose of PAHs in women exposed to air pollution containing notable PAH.

## REFERENCES

Costa DJ, Slott V, Binkova B, et al. Influence of GSTM1 and NAT2 genotypes on the relationship between personal exposure to PAH and biomarkers of internal dose. Biomarkers 1998; 3:411-424

Hawthorne SB, Miller DJ, Barkley RM, Krieger MS. Identification of methoxylated phenols as candidate tracers for atmospheric wood smoke pollution. Environ Sci Technol 1988; 22: 1191-1196.

MMWR. Populations at risk from air pollution—United States, 1991; 42: 301-304, 1993.

NRC. Human exposure assessment for airborne pollutants. National Academy of Sciences, Washington DC, 1991.

Sexton K, Ryan BP. Assessment of human exposure to air pollution: Methods, measurements, and models. IN Air pollution, the automobile, and public health. Eds Watson NY, Bates RR, Kennedy D. National Academy Press, Washington, DC 1988.

Yanagisawa Y, Hishimura H. A badge-type personal sampler for measurement of personal exposure to $NO_2$ and NO in ambient air. Environ Int 1982; 8: 235.

Zhu X. Development of a potential biomarker of environmental wood smoke exposures. Master's Thesis, Department of Environmental Health, University of Washington, Seattle, 1998.

# CHAPTER 6.

# INHALATION EXPOSURE TECHNOLOGIES

Inhalation exposure systems are necessary to conduct controlled experiments of various test atmospheres. Inhalation toxicology experiments involve such controlled exposures either using whole animals, humans, or cultured cells. Inhalation exposure systems need to be designed and used very carefully to achieve the desirable endpoint. Exposure of subjects requires a dependable gas or aerosol delivery system and adequate monitoring instruments to insure proper temperature, humidity, air flow, and atmosphere concentration. Inhalation exposure systems range in size and complexity from simple blow by systems for small animals or cells to whole body chambers for controlled human experiments that can be 20-70 cubic meters in size. Samples of such systems are listed in Table 6-1.

*Table 6-1. Various inhalation exposure systems*

*Blow-by system*
- *Easy to control*
- *Continuous Measure of minute ventilation ($V_E$)*
- *Mouthpiece or mask exposure (Unrealistic, artificial breathing)*

*Whole body chamber*
- *Unencumbered breathing*
- *<u>Very</u> expensive*

*Head Dome*
- *Unencumbered breathing*
- *Inexpensive*
- *With care can measure $V_E$*

---

The characteristics that must be controlled in an exposure system are given in Table 6-2.

*Table 6-2. Characteristics of gas-aerosol exposure systems*

*Temperature and humidity controls for inspired air*

*Gas and particulate scrubbers or filters to cleanse the mixing air*

*Delivery of the atmosphere to the breathing zone at a physiologically adequate flow rate*

*Atmosphere samplers (for temperature, relative humidity, and atmospheres) placed in the breathing zone*

*Safe exhaust of the test atmosphere*

*Systems for increasing the ventilation rate of subjects; treadmill or ergocycle*

*Air cooling system if warranted*

*Calibration instruments for quality control*

---

Several reviews of methods of inhalation exposure systems are available. A recent review of systems for human exposures is recommended if the reader wants more depth on this subject than this text offers (Folinsbee et al, 1997). An earlier review covers animal inhalation exposure systems (Phalen, 1984). This text will give a brief description of the advantages and disadvantages of the more commonly used exposure systems for common air pollutants.

Mouthpiece exposures using blow-by atmosphere delivery have many advantages. The major advantages are that only a small amount of test atmosphere needs to be generated and, with use of a pneumotachograph, minute ventilation can be recorded continuously during exposure allowing precise calculation of inhaled concentration and estimation of inhaled dose. However one obvious disadvantage is the artificial nature of breathing. Also the differential scrubbing effect of nasal versus oral or oronasal breathing has been targeted as a consideration in extrapolating results to actual ambient exposures. This becomes a significant problem with exposures to highly water soluble gases such as $SO_2$. Mouthpiece exposures by pass the nasal defenses. This problem is discussed in the coverage of the health effects of $SO_2$ in Chapter 9. Actually both mouthpiece or face mask (nasal breathing) exposures are artificial since it has been shown that during normal inhalation, even during vigorous exercise, air flows into the lungs through both the mouth and nose. Niinimaa and co-workers estimated that 70-85% of healthy adults breathe oronasally during moderate exercise (1980). Also there is concern that mouth only breathing will dry the airways and may induce exercise-induced bronchoconstriction in certain subjects. Brain and Sweeney (1989) suggest that the loss of filtering capacity resulting from the switch to oral breathing may contribute to the phenomenon of exercise-induced bronchoconstriction.

Whole body chambers remove the disadvantage of the artificial breathing pattern, however chambers have some disadvantages of their own. The major disadvantage is the expense of the accessory instrumentation needed to monitor both the generation and distribution of the test atmosphere. Also chambers are poorly suited to the study of unstable compounds or contaminants which, in very low concentrations, may be diluted by subject re-breathing. Production of ammonia from perspiration during human exercise exposures can neutralize acidic atmospheres and change their chemistry. Finally chamber exposures do not allow easy measurement of minute ventilation.

The facemask has been used both in studies of nasal versus oronasal inhalation with $SO_2$ (Koenig et al, 1983) and in determining the regional deposition of inhaled nitric and sulfuric acid fogs (Bowes et al., 1989). However the nasal mask can change the configuration of the face thus altering normal breathing patterns. Little research has been done on the delivery of contaminants to human subjects in a system allowing oronasal breathing other than whole body chambers (Phalen, 1984).

Thus each of the methods has advantages or disadvantages related to degree of precision in assessing dose, loss of contaminant to the subject or system, ease of operation or cost of initial set-up and routine maintenance costs. Particularly in the case of studies of human health effects the need for optimizing the attainment of research objectives for each exposure session becomes of great importance. The development of a method that minimizes operation costs without sacrificing the validity of data would greatly expand the arena of air pollution research. This leads to a description of a head dome delivery system.

The prototype head dome chamber was developed by Dr. Stephen Bowes and coworkers (1990). The chamber consists of a plastic cylinder of approximately 40 liters which rests on the shoulders, counterweighted by an overhead support allowing relatively free movement of the subject. The method allows a relatively inexpensive alternative to whole body chambers, thereby expanding the scope of inhalation studies in research settings. The head dome in use at the University of Washington consists of a bell-shaped cylinder of 0.32 cm (1/8 inch) clear polycarbonate plastic fitted with inlet and exhaust ports (Figure 6-1). A primary factor in the choice of materials was to minimize the losses of the contaminant gas or aerosol to the generation and delivery system, maximizing the stability of the target experimental concentration. Materials were also chosen to minimize weight and allow visibility to promote subject comfort, thereby extending the wearing time. A 0.32 cm (1/8 inch) thickness spun aluminum mold was fabricated as the initial step in the roto-molding process. The plastic dome was produced by a commercial plastics fabricator using the rotary molding technique. Optical clarity of the final product was accomplished by highly polishing the mold surface. The finished dome is 40.6 cm (16 inches) in height, 35.6 cm (14 inch) in diameter and has a uniform wall thickness of 0.32 cm (1/8 inch) for a final volume of approximately 40 liters.

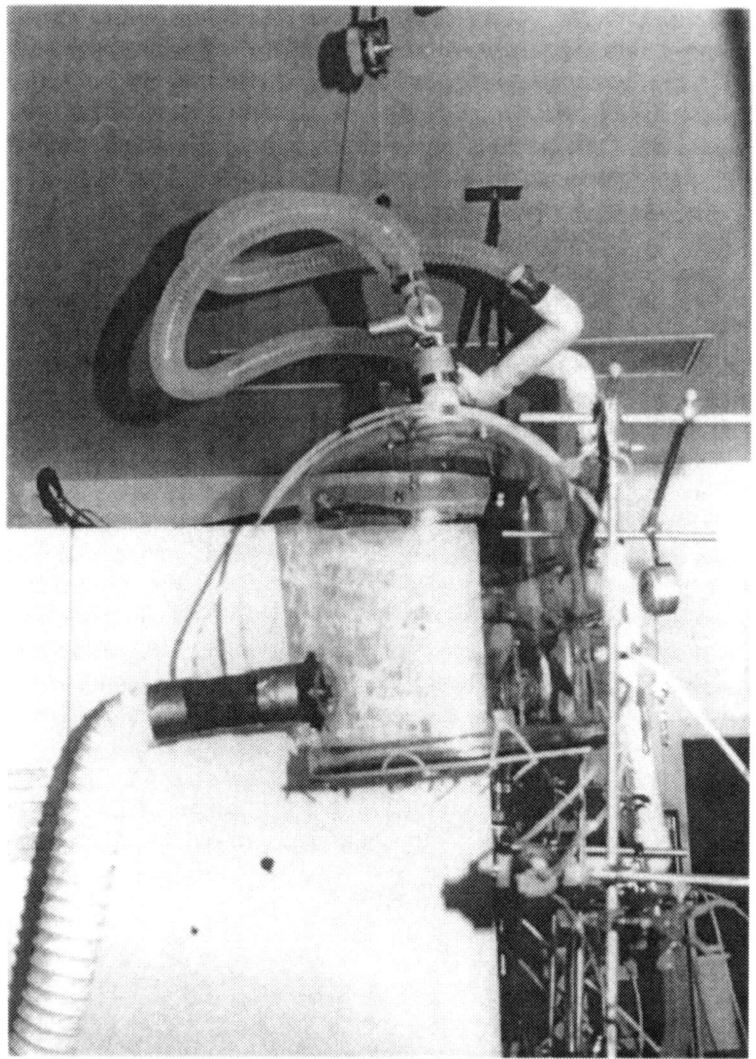

*Figure 6-1. A photograph of the University of Washington head dome. FROM: Kuntz, 1990*

The head dome was then compared with a standard mouthpiece delivery system in an experiment designed to elicit $SO_2$-induced bronchoconstriction. Eight subjects inhaled $SO_2$ on two separate occasions, once delivered by the blow by mouthpiece system and once using the head dome. This comparison found that the head dome system was accepted by the subjects and that the concentration of $SO_2$ was acceptably constant within the dome (Kuntz, 1990). Since that time the head dome has been used in several successful controlled exposure studies.

In vitro exposures to air pollutants require some additional attention. Cultured cells must be kept at approximately 37°C and close to 100% RH. This factor necessitates additional attention to these variables. Cultured cells also need to be in a media that resembles human body fluids. On the other hand, many types of media used for bathing cells will interfere with the delivery of gases such as ozone to the cell surface. Ozone is extremely reactive and will break down into $O_2$ and a free radical instantly upon contact with almost any surface. Researchers conducting in vitro studies of the effects of air pollutants have developed methods to overcome this problem. Cells are grown on porous filter membranes that allow the basal side of the cell to be immersed in culture media while the apical side of cell is in contact with air. This cell-air interface permits the apical side of the cell to be exposed to gas such as ozone, $NO_2$, or $SO_2$. The cells can be placed on an exposure platform situated above the water in a standard heated water bath thus satisfying the requirements for a warm, moist atmosphere. This method of exposure of cells has been used successfully in ozone exposures of respiratory epithelial cells (Beck et al, 1994; Dumler et al, 1994; Potter-Perigo et al, 1998; Jabbour et al, 1998). Figure 6-2 is a schematic of a cell exposure system.

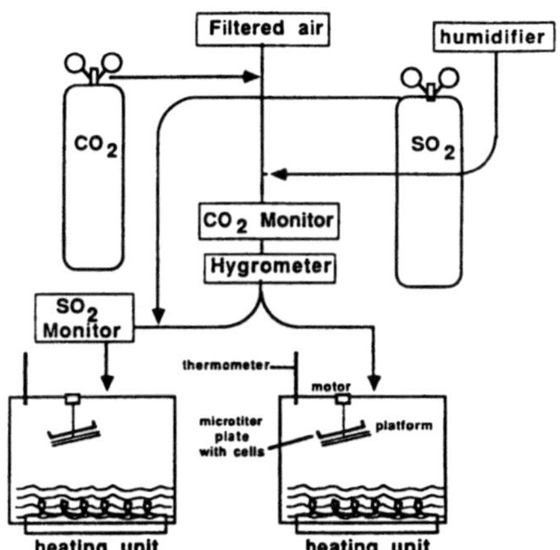


*Figure 6-2. Schematic of a cell exposure system that maintains the cells at > 95% relative humidity and allows simultaneous exposure to a test atmosphere ($SO_2$, ozone, etc) and a clean air control. FROM: McManus et al, 1989 with permission.*

Figure 6-3 is a schematic view of a gas-aerosol generation and delivery system at the University of Washington for use in controlled human studies. This system requires considerably more air flow (800 standard cubic feet/hour) than the in vitro system due to the requirements of human inhalation. This blow-by exposure system has been used with mouthpiece, nasal mask, and head dome exposures. The schematic shows both gaseous and aerosol monitors, as the system can be used to generate and deliver aerosols as well. It has been used for sulfuric acid and ammonium sulfate exposures. This brief review of inhalation exposure systems gives the reader background for the discussions of health effects of air pollutants derived from controlled human studies in subsequent chapters.

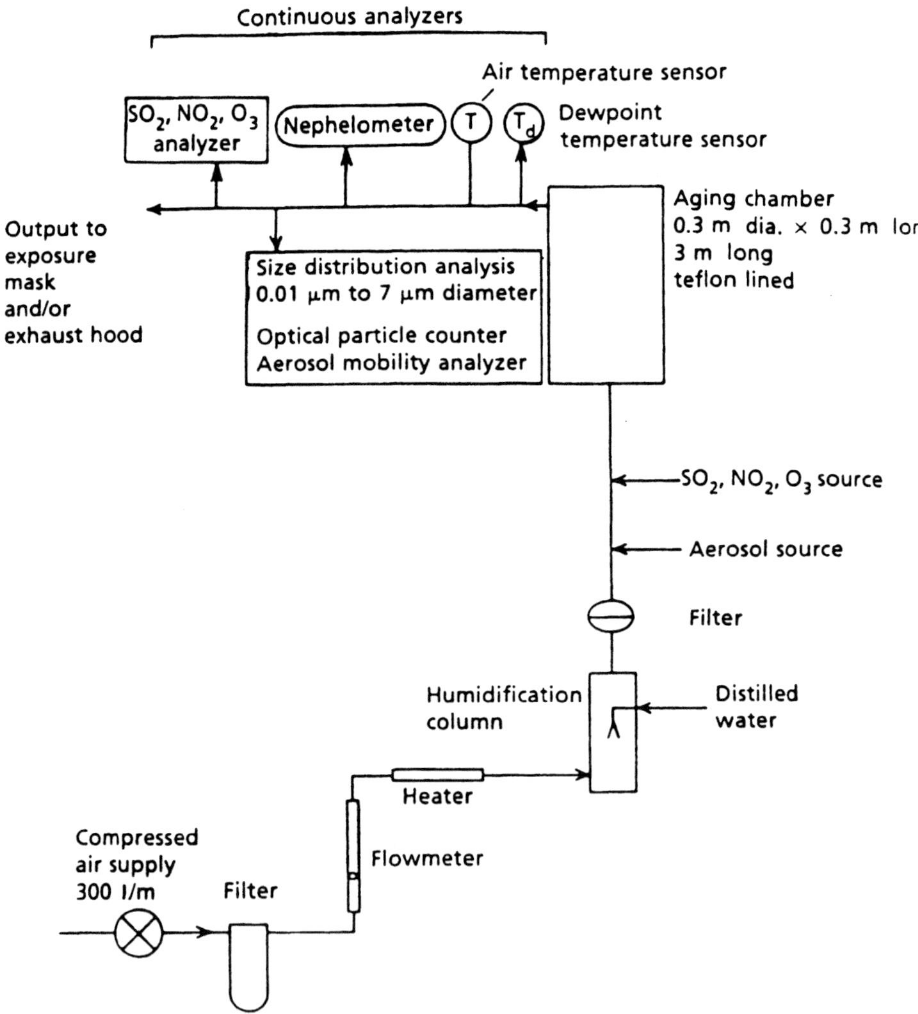


*Figure 6-3. A schematic diagram of a gas-aerosol generation and delivery system for exposure of human subjects.*

# REFERENCES

Beck NB, Koenig JQ, Luchtel DL, et al. Ozone can increase the expression of intercellular adhesion molecule-1 and the synthesis of cytokines by human nasal epithelial cells. Inhal Toxicol 1994; 6: 345-357.

Bowes SM, Frank R, Swift DL. The head dome: a simplified method for human exposures to inhaled air pollutants. Am Ind Hyg Assoc J 1990; 51: 257-260.

Dumler K, Hanley QS, Baker C, et al. The effects of ozone exposure on lactate dehydrogenase release from human and primate respiratory epithelial cells. Toxicol Letters 1994; 70: 203.

Folinsbee LJ, Kim CS, Kerhl HR, et al. Methods in human inhalation toxicology. IN Massaro EJ (ed). Handbook of human toxicology, CRC Press, Boca Raton, 1997. pp 607-672.

Jabbour AJ, Altman LC, Baker C, Luchtel DL. Expression and distribution of _1 integrins in bronchial epithelial cells after in vitro exposure to ozone. Am J Resp Cell Mol Biol 1998; 19: 357-365.

Kuntz K. Characterization of a head dome for human exposures to ambient and occupational airborne contaminants. Masters Thesis, Department of Environmental Health, University of Washington, 1990.

McManus MS, Altman LC, Koenig JQ, et al. Human nasal epithelium: characterization and effects of In Vitro exposure to sulfur dioxide. Exp Lung Res 1989; 15: 849-865.

Niinmaa V, Cole P, Mintz S, Shephard RJ. The switching point from nasal to oronasal breathing. Respir Physiol .1980; 42: 61-71.

Phalen RF. Inhalation studies: foundation and techniques. CRC Press, Boca Raton, 1984.

Potter-Perigo S, Kaplan ED, Luchtel DL, Baker C, Altman LC, Wight TN. Ozone alters the expression of tenascin-C in cultured primate nasal epithelial cells. Am J Resp Cell Mol Biol 1998; 18: 471-478.

# CHAPTER 7.

# METHODS OF ASSESSMENT OF POLLUTANT-INDUCED HEALTH EFFECTS.

There are three primary research fields that provide data of interest to the area of the health effects of air pollution. They are animal toxicology and in vitro systems, controlled laboratory studies, and epidemiology. The three fields of endeavor in air pollution health effects research employ some of the same methods of assessment. For many years almost all the controlled human experiments with air pollutants relied on pulmonary function data as the outcome measure. However, there is increasing use of assessments that may detect more subtle effects of exposures. Also traditional studies of the health effects of air pollution were designed to detect adverse effects in the respiratory system as the lungs were assumed to be the primary target organ. As epidemiologic studies increasingly identified increased morbidity and mortality in persons with pre-existing cardiac disease, studies now are employing methods to detect adverse cardiac effects as well. A list of important assessment tools is given in Table 7-1.

*Table 7-1. Methods of assessment of pollutant-induced effects.*

- *Nasal resistance or work of breathing (see Ch 3)*
- *Lung Volume and capacity measurements (see Ch 3)*
- *Breathing mechanics*
  - *Resistance and/or Compliance (see Ch 3)*
- *Airway reactivity*
  - *Provocational tests*
    - *Pharmacologic: Methacholine, Histamine, $SO_2$*
    - *Physical: Cold air, Exercise, Hypo-osmolar saline*
- *Intrapulmonary gas distribution*
  - *e.g.. Forced oscillatory resistance measurements*
- *Alterations in defense mechanisms*
  - *Particle clearance with Radio-labeled particles*
  - *Deposition patterns (percent retention of inhaled particles)*
  - *Retrieval of alveolar macrophages (AMs) for in vitro assays*
  - *Cilia cytotoxicity*
- *Airway permeability*
  - *$^{99m}$DTPA (diethylene triamine pentacetic acid)*

*Cardiac endpoints*
- *Pulse rate*
- *Arterial oxygen saturation*
- *Heart rate variability*
- *Blood pressure*
- *Cardiac dysrythmias*

*Health endpoint data sets*
- *Emergency room visits*
- *Hospital Admissions*
- *Death Certificates*
- *Physician/clinic visits*
- *Medication use*
- *Symptom and activity pattern diaries*
- *School and work Absenteeism*

---

Each of the three disciplines involved in air pollution research has its advantages and disadvantages for supplying data relevant to an understanding of health effects and are now summarized.

## ANIMAL TOXICOLOGY AND IN VITRO SYSTEMS

Animal toxicology has contributed a great deal to the field of the health effects of air pollution. Many of the animal exposures have served as precursors to controlled human studies by defining the concentration and duration of appropriate exposures. Use of animal studies has allowed us to identify cells within the respiratory system that are susceptible to air pollutants as well as to identify various biochemical mediators of interest for further investigation in human studies. Table 7-2 lists some advantages and disadvantages of animal toxicology studies and studies using in vitro techniques with various types of cell culture preparation.

*Table 7-2. Advantages and disadvantages of animal experiments*

*ADVANTAGES:*
- *Morphological examinations*
- *Biochemical assays*
- *Can use higher doses for mechanistic purposes*
- *Chronic controlled studies*
- *Combination of pollutants*

*DISADVANTAGES:*
- *Extrapolation to humans*
- *Wide inter-species differences*
  - *cell types*
  - *lung morphology*
  - *enzyme systems*
  - *breathing patterns*
  - *age span*

Animal studies are known for their superb control of most variables such as the concentrations of the pollutants, the duration of the exposure, and the species and age of the animal. Investigators can even choose the genetics of the animals they use. In fact, altered genetic material is being used to achieve an even greater control over experiments. Both transgenic mice that are bred for certain characteristics and knock-out mice which have specific genes removed are opening up the field of genetic use of animals. Animals can be used in chronic studies with exposures lasting from weeks to years. Obviously these types of controlled experiments cannot be carried out with human subjects. Investigators can design elegant dose-response studies using higher concentrations of an air pollutant than would be possible in a human study. The endpoints assayed after the exposure can include almost any measurement a researcher can imagine including biochemical assays of lung fluids and cell systems, morphological examination of areas of injury within the respiratory system, and even mortality. The major disadvantage is extrapolation of the data for use in understanding air pollution effects in human subjects. This is compounded by the frequent use of high exposures that do not give data on realistic ambient concentrations of the pollutants. In vitro studies also have the disadvantage of studying cells in isolation away from the normal milieu of interacting cell systems. In summary, animal and in vitro studies are very valuable since they provide the field with mechanistic data. Many scientists and especially the regulatory community want to know the mechanism by which a pollutant causes its reputed effects. The scarcity of data relating to mechanisms by which particulate matter air pollution causes morbidity and mortality has been the major criticism of EPA's action to set a new PM National Ambient Air Quality Standard for PM less than or equal to 2.5 μm. This controversy is discussed in Chapter 11 on the health effects of PM.

## CONTROLLED HUMAN STUDIES

Controlled human studies have provided the field of the health effects of air pollution with some uniquely important data. One advantage of controlled human studies is the high degree of control over important variables. The investigator can select the age, gender, and susceptibility of the subjects. The concentration of the pollutant(s) can be kept at a constant concentration or can be varied at a repeatable level for each subject. The experiment can be conducted with the subject at rest or at various levels of physical activity. The ventilation rate often can be measured so that the inhaled concentration for each subject can be calculated. Individual subjects can be studied at several different pollutant concentrations using a repeated measures experimental design. Repeated measures designs are not used often in animal studies because of the added expense of maintaining the animals and the need to sacrifice the animals for post-exposure analyses. Table 7-3 lists some of the advantages and disadvantages of controlled human studies.

In controlled human studies, the duration of the exposure can be controlled and also, to some extent, the timing of the post-exposure measurements. However making measurements of delayed effects after the subject has left the laboratory is difficult. Some studies send a symptom rating form and peak flow meter home with the subject. However the quality of these data, taken at home without direction and at uncertain time periods, are questionable. Table 7-4 shows a typical symptom rating form.

The greatest advantage of human studies is the lack of any need to extrapolate the findings. However there are many disadvantages to controlled human studies. First and foremost is the exposure duration constraint. Human subjects cannot be exposed for more than about 8 hours per day and not for more than about 5 days in sequence. So human exposures do present extrapolation problems--in this case the problem is to extrapolate from brief acute exposure outcomes to real time daily and yearly exposures. Other disadvantages listed in the table that need emphasis are the inability to study seriously ill individuals and the small sample sizes that these labor and time intensive studies require.

*Table 7-3. Advantages and disadvantages of controlled human experiments*

*ADVANTAGES:*
- *Known subject characteristics*
- *Controlled pollutant concentrations*
- *Controlled activity level*
- *Exact measurements of effects*
- *Randomized and double blind designs*
- *Repeated measures*

*DISADVANTAGES:*
- *Short term exposures*
- *Usually noninvasive measurements*
- *Small sample size*
- *Usually cannot study severely ill persons*
- *Usually study single pollutants*
- *Cannot recreate "real world" pollutant mixtures*

*Table 7-4. A typical symptom rating scale for use in controlled human studies of air pollutant effects.*

*Subject number* _____

*Exposure date* _________*Do you regularly take vitamins?* _____

*Did you drink anything in the last 2 hrs.?*
*When*___________________*What*___________________

*How have you been feeling this week?*
*Better than average* ________*Average*________*Worse than average*________

*Have you had any illness or had an asthmatic attack during the last week?*
_________________________________

*When did you take your last medication?*__________________________________________________

*Please rate the following symptoms dependent on severity:*

| 0 | 1 | 2 | 3 | 4 | 5 |
|---|---|---|---|---|---|
| *None* | | | | *Severe* | |

| | *Before Exposure* | *During exposure* |
|---|---|---|
| *Cough or burning* | ______________ | ______________ |
| *Shortness of breath* | ______________ | ______________ |
| *Fatigue* | ______________ | ______________ |
| *Headache* | ______________ | ______________ |
| *Unusual taste or smell* | ______________ | ______________ |
| *Sore throat* | ______________ | ______________ |
| *Nasal discharge* | ______________ | ______________ |
| *Wheezing* | ______________ | ______________ |
| *Dizziness* | ______________ | ______________ |

## EPIDEMIOLOGICAL STUDIES

The third field of research that provides data in the field of health effects of air pollution is epidemiology. Epidemiology seeks to answer the important questions: that is, what are the effects of air pollution exposure at levels seen in communities on human health. Of course, it is very difficult to study human health in a community and tease out only the adverse health outcomes attributable to air pollution. What about genetic differences, diet, occupational exposures, activity patterns, types of home heating, and of course exposure to either active or passive cigarette smoke? Epidemiology is often criticized for not supplying data that help understand mechanisms of air pollutant effects. However, epidemiology is not designed to answer mechanistic questions. David Bates, one of the founding fathers of air pollution epidemiology, has this to say about the relative advantages of toxicology versus epidemiology.

"...animal toxicologists, who are able to control all variables in their experiments, generally seem to be biased against making causal inferences unless the mechanisms are fully understood. They, and others, may also be influenced by the idea that, since good science is necessarily cautious, being more cautious represents better science. This kind of primitive thinking often passes as wisdom." (Bates, 1994 p 64).

Although epidemiology is not designed to establish causative relationships, epidemiologic studies can imply causation. Epidemiologists often grade the importance of their findings with respect to showing a possible causative association by comparing them to a set of criteria developed years ago by Brandon Hill. These are listed in Table 7-5.

*Table 7-5. Aspects of association necessary to establish causation. Adapted from Hill, 1965.*

| | |
|---|---|
| *Strength of association* | *Is a given exposure to the subject agent associated with a large effect?* |
| *Consistency* | *Has the effect been observed consistently in different studies?* |
| *Specificity* | *Is exposure to the subject agent associated with particular sites and types of diseases?* |
| *Temporality* | *Does exposure to the agent precede the effect ?* |
| *Biological gradient* | *Is there a dose-response relationship?* |
| *Plausibility* | *Is there a plausible biological mechanism to explain the effect?* |
| *Coherence* | *Is the association consistent with the known facts of the natural history and biology of the disease?* |
| *Experiment* | *Does reducing exposure or blocking the action of this agent reduce or prevent the disease?* |
| *Analogy* | *Do other agents, similar to the one in question, produce the effect?* |

The exacting list of criteria given in Table 7-5 were intended for environmental epidemiology in general. However they are extremely pertinent to the field of the epidemiology of air pollution. In Chapter 11, where there is a discussion of the health effects of particulate matter, each of these criteria will be applied to the case of PM-induced health effects to give the reader a basis for reaching his or her own conclusion about the strength of the association between exposure to PM air pollution and various endpoints of morbidity and mortality.

At this point it is appropriate to list the advantages and disadvantages in studies of the epidemiology of air pollutants (see Table 7-6).

*Table 7-6. Advantages and disadvantages of human epidemiologic studies*

*ADVANTAGES:*
*Study chronic effects*
*Realistic exposures*
*Real-time lung function and respiratory symptom measurements*

*DISADVANTAGES:*
*Inadequate air monitoring data*
*Low level of pollutants in US at present*
*Poor characterization of the actual composition of pollutants over time;*
*Inadequate health endpoints based on recall of symptoms or diaries with poor quality assurance.*

---

The chief combined advantage to epidemiologic studies is their ability to study either acute or chronic effects associated with real-time air pollution exposures while making health endpoint measurements after an integrated period of exposure or over an integrated period of time. The chief disadvantage is the inability to establish an exact (or even close estimate) of the population exposure to the pollutants of interest. These advantages and concomitant disadvantages are typified by the time series epidemiologic studies that have provided much of the information on the health effects of PM.

Time series studies are often classified as "ecological" or "observational" studies since they do not intervene or in any way control the variables of interest. A time series air pollution study is a test of associations between a daily time series of an air pollutant and a daily time series of a health outcome (emergency department visits, hospital admissions, or mortality). Time series studies are by their nature longitudinal in design. Several years of daily data are usually required with relatively fewer years needed for necessary statistical power in studies using emergency department data and relatively more in mortality studies. The time series design attempts to control for the effects of other time-varying data that might confound the finding. In general, time series studies do not need to attempt to control for personal habits of individuals since variables such as diet and smoking occur in the same pattern each day and there is no reason to believe the daily patterns of these habits would be influenced by air pollution. On the other hand, the time series design must control for daily meteorological events (temperature, humidity, wind speed) and for other co-varying air pollutants. Since many air pollutants (CO, $NO_2$, and $PM_{10}$) are emitted from the same sources and disperse throughout an air shed in a similar fashion, their 24 hour concentrations are highly correlated and their potential influence on the pollutant-health outcome of interest is great. An example of the high correlations among pollutants is given in Table 7-7. Single pollutant and multiple pollutant models are compared to evaluate the relative importance of co-pollutants, as we will see in the discussions of specific research findings in the chapters on health effects. The results in epidemiologic studies are often reported as a relative risk (RR), which is the ratio of disease or symptom incidence among exposed to that of the unexposed population. The relative risk is placed into context by a 95th percentile confidence interval which is analogous to a p value = to 0.05. The confidence interval can be judged for statistical significance which occurs when the lower value in the interval is greater than 1.0.

Another experimental design used in the epidemiology of air pollution design. Experiments using the cross-sectional design usually compare some health outcome in one city or region with higher air pollution to a city or region with lower air pollution. Although this appears to be a straight-forward design, it has distinct disadvantages due to the fact that the two populations may have significant demographic differences unknown to the investigator. Unlike the time series design, personal patterns of cigarette smoking and diet are crucial in the cross-sectional design and need to be considered and adjusted. Cross sectional studies can use a time-series design or they can use a prospective panel study design where individual subjects are actually recruited and thus personal habits are noted and can be entered into the analysis. However, panel study sample size is severely limited by the intensive nature of the design and thus much statistical power is lost.

Table 7-7. Bi-variate correlation among exposure variables in Seattle (9/1/95 to 12/31/96)

| | $CO^a$ | $PM_{10}^{a,b}$ | $\sigma_{sp}^a$ | $NO_2^c$ | $NO_2^a$ | $SO_2^c$ | $SO_2^a$ |
|---|---|---|---|---|---|---|---|
| $CO^a$ | 1.00 | 0.74 | 0.74 | 0.47 | 0.66 | 0.15 | 0.32 |
| $PM_{10}^{a,b}$ | | 1.00 | 0.82 | 0.56 | 0.66 | 0.24 | 0.43 |
| $\sigma_{sp}^a$ | | | 1.00 | 0.41 | 0.59 | 0.19 | 0.34 |
| $NO_2^c$ | | | | 1.00 | 0.85 | 0.22 | 0.37 |
| $NO_2^a$ | | | | | 1.00 | 0.45 | 0.25 |
| $SO_2^c$ | | | | | | 1.00 | 0.82 |
| $SO_2^a$ | | | | | | | 1.00 |

[a]*Daily average*
[b]*Light scattering, a measure of fine PM*
[c]*Daily maximum 1-hr*
*FROM: Norris et al, 1999*

Thirdly, the cohort design follows a group of people over a period of time, measuring the health endpoints of interest at the beginning and end of the time period. Examples of cohort studies will be discussed in Chapter 11 as they relate to the health effects of PM. Cohort studies often use the prospective panel study design asking subjects to record daily activity patterns and health indices. The best cohort studies also measure the personal, indoor, and outdoor concentrations of the pollutants of interest as described in Chapter 5 on exposure assessment. This, of course, greatly increases the labor and dollar costs of the research. As mentioned, epidemiologic studies of the health effects of air pollution have evaluated both acute (short-term) effects of air pollution as well as chronic (long-term) effects. Some examples of each type of study are given in Table 7-8.

*Table 7-8. Examples of the endpoints evaluated in typical acute and chronic epidemiologic air pollution studies.*

*Acute Effects*
- *Asthma aggravation*
- *Respiratory infections*
- *Transient changes in pulmonary function*
- *Emergency department visits*
- *Hospital admissions*
- *Daily mortality*

*Chronic Effects*
- *Disease prevalence*
- *Lung Growth or Decline*
- *Chronic mortality (survivability)*

---

The advantages and disadvantages of epidemiologic studies have been summarized. Epidemiologic studies must attempt to adjust for many confounding variables that impact on the strength of the conclusions such as those listed in Table 7-9.

*Table 7-9. A list of potential confounding variables in the field of air pollution epidemiology.*

| *Confounding variables* |
|---|
| *Tobacco smoke exposure* |
| *Occupational exposure* |
| *Variation in diets* |
| *Respiratory infections.* |
| *Gender, age, and racial differences* |
| *Pre-existing respiratory or cardiac disease* |

## BIOMARKERS

Other factors that also bear on the difficulty with epidemiologic experimental designs are relative infrequency of some chronic respiratory diseases, multifactorial nature of these chronic diseases, lack of sensitivity and specificity of some of the tools used to detect physiological dysfunction or disease, and the difficulty in estimating biologically effective dose. Some of these difficulties can be attenuated by the use of biomarkers. Biomarkers are biologic indicators of exposure, effects, or susceptibility derived from blood, breath, urine, or other tissue. Biomarkers add objectivity to epidemiologic studies because in most cases they can be quantified. Unfortunately, biomarkers are not yet available for all compounds or pollutants that are studied or of interest. Table 7-10 gives a list of potential sources and identities of some biomarkers used in air pollution studies.

*Table 7-10. Biomarkers of Air Pollutant-Induced Effects in Human Subjects*

*1) Biomarker of exposure--potential sources:*

- *Deposited material*
  - *Blood, urine, teeth, hair*
  - *Respiratory tract fluids: Sputum; Saliva*
  - *Exhaled air*

*2) Biomarkers of effects:*

- *Potential sources of samples:*
  - *Same as above; lavage fluids or cells*

*Potential inflammatory or cellular injury mediators (selected examples, see abbreviation list for identification):*

*Histamine, Adhesive molecules eg ICAM-1, Tumor necrosis factor alpha (TNFα), integrins. Platelet activating factor (PAF), Immunoglobulin E (IgE), Leukotrienes vs Prostagladins, Interleukins (IL) IL1-IL12*

*Potential cellular biomarkers*

*Neutrophils,* | *Eosinophils*
*Mast cells* | *T lymphocytes*
*Alveolar macrophages* | *Epithelial cells*

*Potential markers of injury:*
*Lactose dehydrogenase (LDH)*
*Glutathione*
*Uric acid*

*3) Biomarkers of susceptibility:*

*Age*
*Ethnicity*
*Genetic susceptibilities*
*Cystic Fibrosis*

*Pre-existing disease:*
*Asthma: (IgE; BHR; EIB; Skin prick tests)*
*COPD*
*Idiopathic Pulmonary Fibrosis*

---

Each of the biomarkers listed above has been used to help evaluate the health effects of air pollutant exposure. Examples of biomarkers of exposure are carboxyhemogloblin, a sensitive measure of CO exposure over the past 24 hours and blood lead levels. (An example of a biomarker for wood smoke exposure is given in Ch 5 during the discussion of enhanced exposure assessment.) Examples of biomarkers of effects are neutrophils and lactate dehydrogenase that have been shown to be increased in both nasal and alveolar lavage fluid following exposure of human subjects to ozone. Examples of effects of ozone exposure to cultured cells are increased expression of TNF-_ and ICAM-1. Examples of biomarkers of susceptibility are the presence of documented asthma or COPD and the use of children or older individuals in an attempt to enhance the ability to detect air pollution effects by selecting subjects that may be more susceptible to these agents. Specific examples of studies using these assessments of health effects of various air pollutants are given in the chapters

## POLYMORPHISMS

One of the newest assessment tools for understanding air pollutant-induced effects is measures of genetic variability. Many enzymes and proteins in our bodies have several different forms, that is they are polymorphic. Current molecular biological assays can test the amino acid sequences of proteins and determine whether polymorphisms for a given protein exist. One research goal is to identity polymorphisms that are associated with increased susceptibility to air pollution. As discussed in Ch 8, asthma is a multifactorial disease with many contributing factors. Although environmental factors play a prominent role, there is also a genetic basis to asthma. A number of polymorphisms for proteins that play important roles in asthma have been identified. These genes could play a role in susceptibility to air pollutants. Three potential genes of interest for asthma are the $\beta_2$-adrenergic receptor, the HLA Class II factor, and the $\alpha$ subunit of the IL-4 receptor.

The $\beta_2$-adrenergic receptor is responsible for smooth muscle tone of the bronchial airways, as discussed in Ch 2. Due to this function, pharmaceutical companies have developed medicines for asthma such as albuterol that stimulates the $\beta_2$-adrenergic receptor causing dilation of the airways. $\beta_2$-adrenergic dysfunction may be one of the underlying mechanisms responsible for atopy as well as bronchial asthma (Ohe et al, 1995). The gene encoding the human $\beta_2$-adrenergic receptor ($\beta_2$AR) has recently been isolated and sequenced. There are three polymorphisms of this receptor that are of interest: one at amino acid position 16 where glycine has been substituted for arginine (Arg/Gly 16), one at amino acid position 27 where glutamate has been substituted for glutamine (Gln/Glu 27), and one at amino acid position 146 (Thr/Ile 164). These substitutions impart specific biochemical and pharmacologic phenotypes to the receptor. A recent study demonstrated that the Gly16 allele of the $\beta_2$ AR, which relates an enhanced downregulation of receptor number, is significantly overrepresented in nocturnal asthma (Turki et al, 1995). Other investigators have shown that the $\beta_2$AR Gly16 allele is associated with a subset of asthmatic patients with a distinct clinical profile of severe disease (Reihsaus et al, 1993). Differences in expression of $\beta_2$ adrenergic receptors conferred by the various alleles may influence airway responsiveness to air pollutants in people with asthma.

Human leukocytic antigen (HLA) Class II molecules also have been associated with asthma. The HLA system is a genetic loci closely associated with disease susceptibility. Investigators following soy bean-induced asthma in Barcelona noted that only a small number of subjects with asthma in that city were affected by the soy bean dust (Soriano et al, 1997). This suggested to them that there might be a variation in genetic predisposition to such aggravation. They hypothesized that a genetically determined human leukocyte antigen (HLA) Class II factor might play a role and, in fact, they found that the risk of soybean epidemic asthma was associated with the HLA-DRB1*13 allele ($p = 0.02$). The only individuals with asthma that developed symptoms from inhaling soy bean dust had the HLA-DRB1 form of the protein.

Interleukin-4 (IL-4), a growth factor for mast cells, plays a significant role in the development and status of asthma through B-cell activation and immunoglobulin E (IgE) production (Mencia-Huerta et al, 1990). IgE is a biomarker for asthma as described in Ch 8. Also the interleukin-4 receptor $\alpha$ subunit plays a central role in regulating the production of IgE and a recent study has identified a novel interleukin-4 receptor $\alpha$ allele (Hershey et al, 1997). This genetic polymorphism encodes for a substitution of arginine for glutamate at amino acid position 576 (Glu/Arg 576). Further studies have shown that the variant polymorphism enhances signaling function. In addition, the polymorphism was common among patients with allergic inflammatory disorders, with a relative risk of atopy among those with a mutant allele of 9.3 ($p=0.001$). This mutation may predispose persons to allergic-related diseases such as asthma.

Assays for polymorphisms can be performed by extracting DNA from a tissue sample. Buccal cells from the inside of the cheek are very easy to obtain and are frequently used for polymorphism studies. The sample is collected with a soft brush. The subjects brush the inside of each cheek about 5 times always rotating the brush in the same direction. The brush is then cut away from the handle and stored in sterile distilled water and stored for DNA extraction.

## CONCLUSION

This chapter has covered a long list of methods that have been used to assess the health effects of air pollutants in specific studies. Individual investigators design the individual studies and thus it is often difficult to compare across experiments. There have been attempts to standardize protocols for controlled human studies. Controlled human subject protocols that would be useful for standardization are methacholine testing for BHR in all subjects, statistical tests, post exposure-endpoint intervals, which endpoints to use, duration of exposures, and so forth. Most of the assessments described are objective measurements. However, they eventually need to be interpreted by the authors of the articles and by the scientific audience in general. There has been considerable debate about what respiratory changes should be considered adverse. The definition of adverse effects is discussed in more detail in Ch 18. Also there is debate about the difference between statistical significance and clinical significance. EPA prepared a list showing a gradation of physiologic responses to ozone that may be useful to the reader in evaluating some of the studies to be described throughout this book. That list is given in table 7-11.

*Table 7-11. Gradation of individual physiological responses to acute ozone exposure.*

| Physiologic measure | Gradation of response | | |
|---|---|---|---|
| Mild | Moderate | Severe | Incapacatating |
| *Change in spirometry (eg, $FEV_1$, FVC)* | | | |
| 5-10% | 10-20% | 20-40% | >40% |
| *Duration of effect* | | | |
| < 30 min | < 6 hrs | 24 hrs | >24 hr |
| *Symptoms* | | | |
| None | few mild cough | Repeated cough, inspiratory pain shortness of | Severe cough distress |
| *Limitation of activity* | | | |
| None stop | some subjects stop | more stop | most stop |

The purpose of this chapter has been to acquaint the reader with the array of assessment tools available in the study of the health effects of air pollution. It is hoped that understanding these methods of assessment will assist in understanding the studies that are described in the next six chapters.

## REFERENCES

Bates DV. Environmental health risks and public policy: Decision making in free societies. University of Washington Press, Seattle, 1994.

Hershey GKK, Friedrich MF, Esswein L, et al. The association of atopy with a gain-of-function mutation in the α subunit of the interleukin-4 receptor. N Engl J Med 1997; 337: 1720-1725.

Hill AB. The environment and disease: association or causation? Proc Royal Soc Med 1965; 58: 295-300.

Mencia-Huerta JM, Dugas B, Braquet P. Immunologic reactions in asthma. Immunol Allergy Clin N Amer 1990; 10: 337-353.

Norris G. YoungPong SN, KoenigJQ, et al. An association between fine particles and asthma emergency department visits for children in Seattle. Environ Health Perspect 1999; 107: 489-493.

Ohe M, Munakata M, Hizawa N, et al. Beta2 adrenergic receptor gene restriction fragment length polymorphism and bronchial asthma. Thorax 1995; 50:353-359.

Reihsaus E, Innis M, MacIntyre N, Liggett SB. Mutations in the gene encoding for the β2-adrenergic receptor in normal and asthma subjects. Am J Respir Cell Mol Biol 1993; 8: 334-339.

Soriano JB, Ercilla G, Sunyer J, et al. HLA class II genes in soybean epidemic asthma patients. Am J Respir Crit Care Med 156: 1394-1398, 1997.

Turki J, Pak J, Green SA, et al. Genetic polymorphisms of the β2-adrenergic receptor in nocturnal and nonnocturnal asthma. J Clin Invest. 1995; 95:1635-1641.

# CHAPTER 8.

# ASTHMA: A SPECIAL CASE OF SUSCEPTIBILITY TO AIR POLLUTION

as the fish that thrusts its jaw to water draw
so I some air do seek to snare,
and like the weanling goat upon the nipple I suck,
but it is not milk so sweet I crave, but air, air so pure

FROM: Night Attack by (Horowitz,1996)

## DEFINITION AND DIAGNOSES

Although asthma has been recognized as a pathological respiratory condition for over 2000 years, there is no agreed upon definition of asthma. In the 1950s the hallmark of asthma was considered to be reversible airways obstruction. Reversibility of decreased lung function by bronchodilating medications was used to diagnose asthma. Currently, airway inflammation is considered to be the hallmark of asthma. Partly this came about with the introduction of the fiber optic bronchoscope that allowed investigators to examine the airways of patients with asthma. These bronchoscopic examinations revealed intense inflammatory lesions (Barbee and Murphy, 1998). In a recent text on asthma, Woolcock (1997) admits that "in spite of the immense amount of material presented here we do not know the answer" to the question, What is asthma? In the face of this deficit, investigators are relegated to describing symptoms, exacerbations, and abnormal structural components of the airways. There is a current working definition of asthma

" a chronic inflammatory disorder of the airways in which many cells and cellular elements play a role, in particular, mast cells, eosinophils, T lymphocytes, neutrophils, and epithelial cells. In susceptible individuals this inflammation causes recurrent episodes of wheezing, breathlessness, chest tightness, and cough, particularly at night and in the early morning. These episodes are usually associated with widespread but variable airflow obstruction that is often reversible either spontaneously or with treatment. The inflammation also causes an associated increase in the existing bronchial hyperresponsiveness to a variety of stimuli." (NAEPP, 1997).

Currently the diagnosis of asthma is made using a combination of patient history, physical examination, and objective monitoring of pulmonary function over time. As well as not having an absolute definition of What is asthma, we also do not know what causes asthma.

Since asthma cannot be precisely defined, what do we know about asthma? Asthma is the most common chronic illness of childhood and is the major cause of school absenteeism. People with asthma experience more than 100 million days of restricted activity and 470,000 hospitalizations annually in the US. Also in the US more than 5000 individuals die from asthma annually. Hospitalizations are most common in children and in African-Americans, a group most at risk for death from asthma. The prevalence of asthma among children varies from 10 -18% depending on the study. Epidemiologic studies show that the presence of wheezing illness (believed to be a sign of asthma) is increasing in most countries.

The recognition of asthma as primarily an inflammatory disorder has stimulated research into identification of mediators that are discussed later in the chapter. Oxidant damage is known to be partially responsible for the underlying inflammation in asthma (Greene, 1995). In addition, the presence of chronic inflammation in asthma creates oxidative stress and alters the redox status in the lung. The role of environmental factors, specifically air pollution (both indoor and outdoor), in the increasing asthma prevalence and mortality from asthma has recently received considerable attention in the literature (Seaton et al, 1994; Peat, 1994; Pierson and Koenig, 1992; Gergen and Weiss, 1990).

For several reasons, asthma is a syndrome of special interest for understanding the health effects of air pollution. Asthma, by its nature, is a condition that is irritated or aggravated by inhaled agents such as the common air pollutants discussed in this text. Thus, it is natural to examine the effects of various air pollutants in individuals with asthma. Also asthma is a disease which often is manifest in young individuals. Since childhood itself may be a risk factor for adverse response to inhaled agents, the combination of infancy and early childhood and asthma team up to cause considerable concern regarding air pollution exposures. Due to these reasons a very brief summary of the various markers and mediators of asthma is included in this book. Detailed descriptions can be found in many sources such as a recent two volume text titled Asthma (Barnes et al, 1997).

## NATURAL HISTORY OF ASTHMA

According to Barbee and Murphy (1998), determination of the natural history of a disease depends on three criteria. 1) a precise definition of the disease; 2) longitudinal data of the disease from onset, through either remission or death; and 3) a clear understanding of the effect that therapy has on the course of the disease. A study of asthma does not come out well judged by these three criteria. The lack of a consensus definition of asthma has been stated. There also are not many good longitudinal studies and new asthma medications have been introduced so frequently that there in no consensus on the effect of one specific therapy on the course of asthma. In spite of the dismal picture painted above, we do know many of the risk factors for the development of asthma. These are listed in Table 8-1.

*Table 8-1. Risk factors for the development of asthma in childhood.*

*Atopic status*
*Familial history, especially maternal history*
*Gender*
*Premature birth*
*Early respiratory infections*
*Outdoor air pollution*
*Indoor air pollution*
*Environmental tobacco smoke exposure (in utero/ during infancy)*

---

During childhood, asthma is approximately twice as common in boys than in girls. This 2:1 ratio changes at puberty. The natural history of asthma shows that development of asthma is primarily dependent on events early in life. It appears that environmental exposure to potent allergens such as dust mites or ETS during developmentally sensitive periods of life can predispose toward the development of a specific form of the T lymphocyte. That is, some environmental events may determine which specific class of T lymphocytes a child develops, T helper 1 or 2 cells ($T_H1$ or $T_H2$). The presence of $T_H2$ cells and IL-4 appear necessary for the presence of asthma. The role of infections in development of asthma is a controversial topic as there are some data that show that certain childhood infections can protect against asthma. This contention is supported by the relationship between birth order and the presence of asthma. Asthma is more prevalent in the first born, and this prevalence is directly related to the order of birth. The hypothesis is that children in a family cross infect each other and the increased chance of infections seems to be negatively related to the risk of developing asthma. The biologic basis for this phenomenon involves the role of the T cells. It is believed that certain infections are associated with T cell development by upregulating the $T_H1$ pathway and depressing the $T_H2$

pathway (the pathway which is necessary for the development of asthma).

From an evolutionary point of view, the inflammatory network of the lung may have survived as a defense against parasitic infections, and asthma may be a hold over from historic conditions where most individuals were infected with parasites. In those situations, a full-fledged inflammatory response was defensive. In individuals without such infections, asthma is an inflammatory over-reaction to such minimal health threats as dust mites, cockroaches, cats, and mold.

## ATOPY

Asthma is one of the atopic diseases. Atopy (the word comes from Greek meaning unusual, singular or out of place) is a constitutional predisposition, based on heredity, to acquire certain allergic states. Atopy can be manifested differently in individuals. Three common atopic diseases are:

atopic asthma, usually referred to as allergic asthma
atopic rhinitis, again usually allergic
atopic dermatitis

Major symptoms of asthma are wheezing, shortness of breath, chest pain, cough, and nocturnal waking. Asthma also can be present in individuals who are not atopic. There also is a form of asthma (adult onset asthma) which does not appear to be dependent on early childhood events.

## MORBIDITY AND MORTALITY RATES

Asthma morbidity and mortality are increasing in the US (MMWR, 1995; Weiss and Wagener, 1990). The Centers for Disease Control and Prevention statistics show that the overall annual age-adjusted prevalence rate of self-reported asthma increased 42%, from 34.7 per 1000 to 49.4 per 1000 between 1982 and 1992 (MMWR, 1995). The estimated annual prevalence rates per 1,000 population of self reported asthma for ages 5-34 has the highest prevalence of asthma. Rates for the period 1980 through 1994 are shown in Figure 8-1.

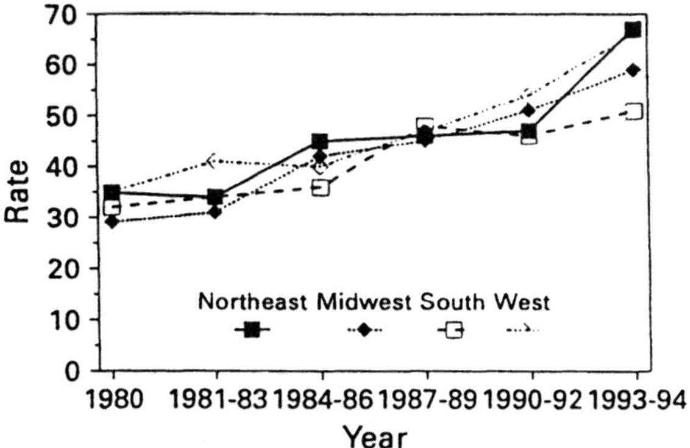


*Figure 8-1. Prevalence of asthma showing an increase from 1980 to 1994. FROM: National Health Interview Survey, 1980-1994, MMWR, 1995.*

## BIOMARKERS OF INFLAMMATION: CELLS

Figure 8-2 is a schematic showing the relationship among various cells important in asthmatic reactions.

### Mast Cells and IgE

Mast cells release chemicals that act immediately on smooth muscle and glands, and also chemicals that attract other cells into the airways. One such mediator is immunoglobulin E (IgE), a protein with known antibody activity. IgE acts to bind antigens to mast cells that then degranulate and spill various mediators of inflammation into the airways. Atopic individuals typically have serum IgE concentrations > 120 IU (international units). IgE is produced by B lymphocytes and is regulated in part by interleukin-4 (IL-4) and interferon-gamma (INF-γ) two mediators of inflammation discussed later. Thus when investigators are evaluating whether a given air pollutant exposure causes inflammation, two biomarkers often assayed are IL-4 and INF-γ.

### Basophils

Basophils and mast cells play a dual role in asthma, as immunoregulators and also as effectors of allergic inflammation. These cells release numerous inflammatory mediators such as histamine, leukotrienes, PAF, trypase, and chymase. These mediators cause asthma-like symptoms. Therefore one could measure these mediators after air pollutant exposure to determine whether the exposure aggravated the asthmatic state.

## Eosinophils

Eosinophils may be the major effector cell in asthma. It is known that serum eosinophilia is a biomarker for asthma, a biomarker of susceptibility as described in Ch 7. Peripheral blood eosinophil counts from patients with asthma are sometimes three standard deviations above normal. Eosinophil counts correlate with $FEV_1$ and PEF values. There is evidence that activated eosinophils produce airway damage. Two proteins associated with eosinophils, eosinophil cationic protein (ECP) and major basic protein (MBP), cause cilia dysfunction and epithelial sloughing. Denuded airway epithelium is a pathologic sign of asthma. Eosinophils also can produce lipid mediators such as leukotriene C4 ($LT_{C4}$) which may cause bronchoconstriction.

## Epithelial Cells

Another cell type, the airway epithelial cell, was originally believed to be primarily a structural type of cell that supported cilia and secreted mucus. It is now recognized that the epithelial cell is a major effector cell in asthma. Studies of bronchial epithelial cells have demonstrated that these cells from subjects with asthma synthesize greater quantities of several cytokines such as IL-1_, IL-16, IL-8, and ICAM-1 (see list of abbreviations). Thus these cytokines are candidates for biomarkers of asthma or asthma aggravation that can be used in studies of the effects of air pollutants on asthma. The epithelium also may contribute to host defense by means of interactions with other inflammatory cells. Epithelial cells appear to play a role in immune reactions.

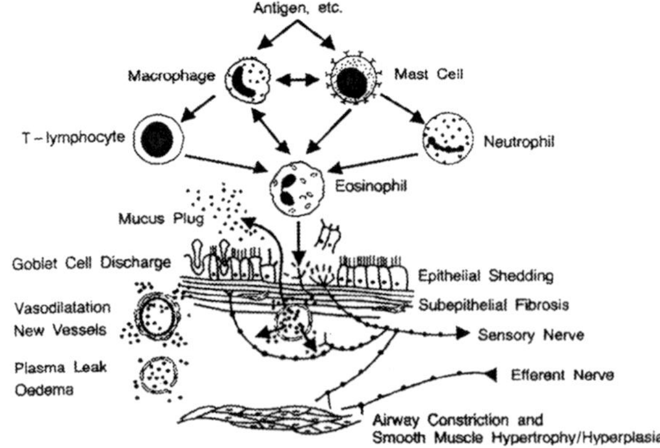


*Figure 8-2. Interactions among various cells and some consequent effects. FROM: Woolcock, 1997 with permission.*

## BIOMARKERS OF INFLAMMATION: MEDIATORS

For readers who wish detailed information on mediators of asthma, there is a recent update by Barnes and associates (1998 ).

### Histamine

Histamine was the first mediator implicated in the pathophysiology of asthma. This mediator is synthesized and released by mast cells and also by circulating basophils. There are two primary receptors for histamine—the H1 and the H2 receptor. The H1 receptor appears to be most involved in mechanisms that lead to the aggravation of asthma. Bronchoconstriction was the first effect of histamine noted regarding asthma. In fact, histamine challenges are sometimes used to diagnosis for asthma, although methacholine is used for that purpose more frequently (see Ch 7). In spite of the potent effect that histamine has on bronchial smooth muscle, anti-histamine medications have not proven very useful in the treatment and management of asthma.

### Leukotrienes

Leukotrienes are products of arachadonic acid metabolism. As a family, they are among the most potent bronchoconstrictors known. In recognition of this, the newest class of asthma medications are agents which influence leukotriene synthesis or action. These drugs can be considered leukotriene modifiers; that is, agents which modify the action of leukotrienes in the airways. At present there are two classes; those that inhibit the action at the leukotriene receptor (montelukast, pranlukast, and zafirlukast) and those that inhibit the action of the enzyme 5-lipoxygenase necessary in the synthesis of leukotrienes (such as zileuton). The leukotriene modifiers are the most specific anti-asthma drugs on the market in that they are designed to act at a precise point in the inflammatory cascade.

### Prostaglandins

Prostaglandins also are products of arachadonic acid metabolism, being synthesized along a different pathway than the leukotrienes. Arachadonic acid was known by the term “slow-reacting substance of anaphylaxis” prior to identification of its chemical form. Some prostaglandins cause bronchoconstriction while others cause bronchodilation.

## Interleukins (IL)

Interleukins play an active role in asthma. They are released from a variety of cells such as mast cells, eosinophils, epithelial cells, macrophages, smooth muscle cells, neutrophils, and fibroblasts. They have been numbered sequentially in the order of their discovery. IL-2 and Il-4 play determining roles in T cell differentiation. Il-8 is known to be a chemoattractant for neutrophils and often is assayed as a biomarker of inflammation in both nasal and bronchial alveolar lavage fluid. Some interleukins are categorized as proinflammatory, whereas others appear to play an inhibitory role in inflammation.

## Tumor necrosis factor-alpha (TNF-$\alpha$)

Tumor necrosis factor- $\alpha$ is primarily released by macrophages and secondarily by lymphocytes. TNF- $\alpha$ is believed to play a role in the late asthmatic reaction. In several studies of ozone exposures, TNF-$\alpha$ increased in nasal epithelial cells following exposure.

## Interferon-gamma (IFN-$\gamma$)

Originally IFN-$\gamma$ was identified as a substance synthesized and released from T lymphocytes. This substance seems to have diverse effects on various cells. From the point of view of asthma, one important effect is inhibition of $T_H2$ cells. However, IFN-$\gamma$ appears also to have proinflammatory effects. Reduced levels of IFN-$\gamma$ are found in subjects with asthma and there is a negative correlation between its levels and the severity of asthma.

## Platelet activating factor (PAF)

PAF is another substance that has long been implicated in the pathophysiology of asthma. It is released from IgE-stimulated basophils and causes inflammation at a variety of sites (platelets, neutrophils, basophils, alveolar macrophages, and eosinophils). PAF is associated with bronchoconstriction, the induction of BHR, and mucus hypersecretion, three of the hallmarks of asthma.

There are numerous proteins classified as cytokines or chemokines that are thought to play an important role in asthma. Only a handful have been mentioned. Investigators are certain to identify new, important molecules in the future. Various growth factors are involved in mechanisms underlying asthma.

## DIET AND ASTHMA

There is evidence that plasma (and presumably epithelial lining fluid) levels of antioxidants such as vitamins C and E influence the asthmatic state. Individuals with asthma tend to have lower plasma concentrations of these antioxidants. This relationship is discussed in more detail in Ch 11 during the discussion of the health effects of ozone. There are several other dietary factors that may influence asthma. Omega-3 fatty acids are one dietary factor frequently shown to be important in asthma. Supplementation with fatty acids was associated with decreases in neutrophil levels (Weiss, 1997). In general, there is a relationship between dietary fish intake (high in fatty acids) and pulmonary function. Since cations such as sodium and potassium influence activity across biological membranes, they have been studied for a role in asthma. There is suggestive evidence that improvement in pulmonary function is associated with treatment with these cations (Weiss, 1997). The study of the interactions between diet and asthma, and also between diet and respiratory response to air pollution, is in its infancy. New, important relationships may well be forthcoming.

## GENETICS OF ASTHMA

Asthma is a complex genetic disorder. It is not a simple autosomal dominant, recessive or sex-linked mode of inheritance. Although one hears a lot about "finding the asthma gene", there ultimately will be several genes associated with different aspects of asthma. Diseases that are largely determined by a single major gene are said to be "simple" in that they approximately obey the classical Mendelian laws or recessive, dominant, or co-dominant inheritance. By contrast, most common diseases (those with greater than 1% prevalence in the population) such as asthma are "complex', in that they are multifactorial or exhibit genetic heterogeneity, and usually both. These 'complex' diseases usually are determined by interactions among major and minor genes and often are influenced by environmental factors (Barnes and Marsh, 1998b ).

The use of polymorphisms as assessment tools for associations between air pollutants and asthma is covered in Ch 7. A brief summary will be given here. Among the genes involved in asthma, the $\beta_2$-adrenergic receptor ($\beta_2$-AR) exists in multiple allelic forms and differences conferred by the various alleles which may influence airway responsiveness in people with asthma. Pharmaceutical companies have taken advantage of the knowledge that the function of the $\beta_2$–adrenergic receptor is dilation of the smooth muscle of the bronchial airways. Beta-adrenergic agonist medications (eg albuterol) are the most widely used inhaled bronchodilators. Two other potential gene polymorphisms for susceptibility to air pollution are in the human leukocyte antigen (HLA) and the IL-4 receptor α allele. In Barcelona, only asthmatics with the DRB1*13 gene of the human leukocyte antigen (HLA) showed an adverse reaction to soybean dust (Soriano et al, 1997). A recent study (Hershey et al, 1997) has identified a novel interleukin-4 receptor α allele which may play a role in allergic inflammatory disorders.

## ECONOMIC BURDEN OF ASTHMA

Weiss et al (1992) estimated the direct medical expenditure and indirect costs (in 1985 dollars) associated with the treatment of asthma and projected these to 1990 dollars. In that report, the annual cost of illness related to asthma was estimated to be $6.2 billion. Inpatient hospital services represented the largest single direct medical expenditure for asthma. Since health care costs are one of the most rapidly growing expenses in the US, we can assume that the cost is now well over $10 billion.

## FATAL ASTHMA

Fatal asthma is rare; there are a little over 5000 cases per year in the US. This low death rate makes it difficult to study the phenomenon. Although the number of deaths is low, it is still unacceptable since asthma should not be a fatal disease. Table 8-2 gives a list of some of the most important risk factors. The relative ranking of risk factors for fatal asthma is not generally known. A recent workshop on Fatal Asthma identified several potential risk factors (Sheffer, 1998).

*Table 8-2. Potential risk factors for fatal asthma.*

*Previous hospitalization for asthma exacerbation*
*Poor compliance with follow-up appointments*
*Liable (unstable) asthma*
*Slowness in seeking medical attention*
*Genetic factors such as response to beta-adrenergic medications*
*Poor perception of asthma symptoms*
*Decreased hypoxic ventilatory response*
*Degree of airway remodeling*
*Resistance to therapeutic effects of inhaled corticosteroid medications*
*Peak flow rate less than 100 l/min*
*Presence of atopy?*
*Socioeconomic status?*
*Indoor environments with high levels of dust mite or cockroach allergens?*

From a morphologic point of view, there are some pathological signs associated with fatal asthma (Table 8-3). These cannot be identified until autopsy and thus are not useful in identifying individuals for increased attention. Also there is a category of individuals with asthma who die very suddenly. These people seem to have an exaggerated neutrophilic response. On the other hand, eosinophilia is more likely to be associated with slow onset fatal asthma.

*Table 8-3. Pathological findings in cases of fatal asthma*

*Hypertrophy of smooth muscle lining of airways*
*Thickened basement membranes, with collagen deposition*
*Mucus plugs*
*Denudation of airway epithelium*
*Resident cells (AMs, masts cells, epithelial cells, endothelial cells--also PMNs, basophils, and platelets*

---

Does air pollution play a role in fatal asthma? This is an extremely difficult question to answer. Due to the rarity of fatal asthma, there is not sufficient statistical power to use traditional epidemiologic protocols to address this question. There certainly are anecdotal examples of individuals who die following an exposure to smoke from grass seed field burning or during a major air stagnation. However, it is usually impossible to state with certainty the cause of non-accidental death. More research may enlighten the interactions between asthma death and air pollution patterns.

Specific studies of associations between common air pollutants and asthma are given in the following chapters. A few examples are cited here. Sulfur dioxide exposures at low concentrations for short time periods (2.5 to 10 minutes) can cause significant airway narrowing in subjects with asthma. Such exposures have been associated with an average decrease in $FEV_1$ of 26%. Nitrogen dioxide exposures in subjects with asthma have been associated with increases in BHR. Subjects with asthma have larger pulmonary function decrements after exposure to ozone than subjects without asthma. These ozone studies have also found that subjects with asthma are the most likely subjects to be susceptible to ozone exposures. Daily PM concentrations have been repeatedly associated with both emergency department visits and hospital admissions for asthma. In the following chapters much attention is paid to other studies that investigate the effects of air pollution on various aspects of asthma.

# REFERENCES

Barbee RA, Murphy S. The natural history of asthma. J Allergy Clin Immunol 1998; 102: S65-S72.

Barnes PJ, Grunstein MM, Leff AR, Woolcock AJ. Asthma Vol 1 & 2. Lippincott-Raven, Philadelphia, 1997.

Barnes PJ, Fan Chung K, Pate CP. Inflammatory mediators of asthma: An update. Pharmacol Rev 1998a; 50: 515-572. Barnes KN, Marsh DG. The genetics andcomplexity of allergy and asthma. Immunol today. 1998b; 19: 325-332.

Gergen PJ, Weiss KB. Changing patterns of asthma hospitalization among children: 1979-1987. JAMA 1990; 264:1688-1692.

Greene LS. Asthma and oxidant stress: nutritional, environmental, and genetic risk factors. J Am College Nutr 1995; 14:317-324.

Hershey GKK, Friedrich MF, Esswein L, et al. The association of atopy with a gain-of-function mutation in the _ subunit of the interleukin-4 receptor. N Engl J Med 1997; 337: 1720-1725.

Horowitz HW. Night attack. Lancet 1996; 348: 252.

MMWR Asthma--United States, 1982-1992. MMWR 1995; 43:952-955.

NAEPP. National asthma education and prevention program. National Institutes of Health. NIH. Publication No. 97-4051A. May 1997.

Peat JK. The rising trend in allergic illness: which environmental factors are important? Clin Exp Allergy 1994; 24: 797-800.

PiersonWE, Koenig JQ. Respiratory effects of air pollution on allergic disease. J Allergy Clin Immunol 1992; 90: 557-566.

Seaton A, Godden DJ, Brown K. Increase in asthma: a more toxic environment or a more susceptible population? Thorax 1994; 49:171-174.

Sheffer AL. Fatal Asthma. Lung biology in health and disease. 1998: 115: 607 pp

Soriano JB, Ercilla G, Sunyer J, et al. HLA class II genes in soybean epidemic asthma patients. Am J Respir Crit Care Med 1997; 156: 1394-1398.

Weiss KB, Wagener DK. Changing patterns of asthma mortality: Identifying target populations at high risk. JAMA 1990; 264: 1683-1687.

Weiss KB, Gergen PJ, Hodgdon TA. An economic evaluation of asthma in the United States. N Engl J Med 1992; 326: 862-866.

Weiss ST, Diet. IN Asthma. Barnes, Grunstein, Leff, Woolcock (eds) Lippincott-Raven, Philadelphia, 1997. pp 105-119.

Woolcock AJ. Overview IN Barnes, Grunstein, Leff, Woolcock (eds) Asthma Vol 1 Lippincott-Raven, Philadelphia, 1997. pp 3-8.

# CHAPTER 9.

# HEALTH EFFECTS OF SULFUR OXIDES: SULFUR DIOXIDE AND SULFURIC ACID

Sulfur dioxide ($SO_2$) is a water soluble gas commonly emitted into ambient air by coal fired power plants, refineries, smelters, paper and pulp mills, and food processing plants. As you recall from Chapter 4, $SO_2$ is part of sulfurous air pollution that has plagued human populations for centuries. $SO_2$ was one of the primary pollutants elevated during the 1952 killer fog in London (discussed in detail in Ch 11). And $SO_2$ is a precursor for sulfuric acid, an air pollutant that plays a major role in the adverse respiratory effects of air pollution. This chapter summarizes what is known about the health effects of both $SO_2$ and sulfuric acid through a review of controlled and epidemiologic studies.

## $SO_2$: CONTROLLED HUMAN STUDIES

### Pulmonary Function

When the Clean Air Act was passed in 1970, scientists intensified the examination of health effects of the criteria pollutants. A dose response relationship between $SO_2$ concentration and pulmonary function decrements was established in healthy adult subjects. In 1980, Koenig and co-workers (1980) first reported that subjects with asthma were especially sensitive to the inhaled effects of acute exposure to low concentrations of $SO_2$. This study is described in some detail as it can serve as a prototype for controlled studies of other pollutants described in later chapters. The study was conducted with nine adolescent subjects (aged 14 to 18 years) who all had allergic asthma and also exercise-induced bronchospasm (EIB). They all required medication to control their asthma symptoms and were categorized as moderate asthmatics. Subjects were exposed to three distinct test atmospheres for 60 minutes at rest on three separate days: clean air, 1 ppm $SO_2$ combined with an inert sodium chloride (NaCl) aerosol, or NaCl aerosol alone. The exposure system used is shown in Figure 6-3. The $SO_2$-NaCl mixture was used to mimic the form most $SO_2$ takes in ambient air (that is, an $SO_2$ aerosol).

Pulmonary function was measured before, during and after the exposures. The measurements recorded were forced vital capacity (FVC), forced expiratory volume in one second ($FEV_1$), total respiratory resistance ($R_T$), and maximal flow rate ($V_{max}$) at both 50% and 75% of expired FVC. Statistical analysis of the pulmonary function data showed a significant decrease in both $V_{max50}$ and $V_{max75}$ when the subjects inhaled the $SO_2$ atmosphere. The percent decreases were 12% and 14% respectively. Since $SO_2$ exposures occur mainly outdoors where young people are active, a second experiment was conducted with similar subjects and test atmospheres but with the addition of exercise on a treadmill during 10 minutes of the exposure (Koenig et al, 1981). The treadmill was set for each subject to increase his or her resting ventilatory rate 5-6 fold (see Figure 9-1). The combination of exercise and $SO_2$ elicited large changes in all the pulmonary functions measures as shown in Table 9-1 below:

*Table 9-1. Percentage change in pulmonary function measurements after exposure to $SO_2$ or air in nine adolescent asthmatic subjects.*

| *Measurement* | *Change from baseline:* with $SO_2$ | with air |
|---|---|---|
| $FEV_1$ | 23% decrease | 0% change |
| $R_T$ | 67% increase | 13%decrease |
| $V_{max50}$ | 44% decrease | 9% increase |
| $V_{max75}$ | 50% decrease | 24% increase |

These changes after $SO_2$ exposure were statistically significant. No significant changes were seen after exposure to air or NaCl alone.

The conclusion from these two studies was that subjects with asthma are extremely sensitive to inhaled $SO_2$ and therefore may be at risk for adverse respiratory effects in communities where $SO_2$ concentrations are elevated even for short periods of time. Other investigators have shown similar results with adult subjects with asthma (Sheppard et al., 1980). Similar studies with healthy subjects often do not find significant pulmonary function decrements after 5.0 ppm $SO_2$.

Further early research provided an interesting result: even allergic subjects who had never been diagnosed as having asthma were extremely sensitive to inhaled $SO_2$ if the subjects had shortness of breath upon exertion (EIB). The effects in this group of subjects were shown in a study similar to the study of adolescents with asthma. Eight atopic adolescents without clinical asthma were exposed to air, a NaCl aerosol, or 1.0 ppm $SO_2$ plus the NaCl aerosol (see Table 9-2). A summary of the

$SO_2$-induced pulmonary function changes can be compared with those given above. In both studies, the changes in pulmonary function are sufficiently large to be clinically important.

*Table 9-2. Percentage change in pulmonary function measurements after exposure to $SO_2$ or air in eight allergic non asthmatic adolescent subjects.*

| *Measurement* | *Change from baseline: with $SO_2$* | *with air* |
|---|---|---|
| *$FEV_1$* | *18% decrease* | *1% increase* |
| *$R_T$* | *41% increase* | *2% decrease* |
| *$V_{max50}$* | *29% decrease* | *4% increase* |
| *$V_{max75}$* | *44% decrease* | *2% decrease* |

The changes after the $SO_2$ exposures were significant.

After these early results, investigators carried out many further experiments designed to characterize $SO_2$-induced pulmonary changes in subjects with asthma. These studies evaluated additive or synergistic effects of $SO_2$ and cold air, the duration of exposure necessary to elicit significant decrements in lung volumes or increases in airway resistance, effect of route of exposure, $SO_2$-induced inflammatory responses, and interactions between $SO_2$ and the effects of various therapeutic agents used to treat asthma. These studies now will be described in more detail.

**Cold Air**

Linn and co-workers compared the response to inhaled $SO_2$combined with either cold air (5°C) or room temperature air (25°C) in adult subjects with asthma (1984). The 5°C temperatures did not exaggerate the response to $SO_2$. However, another group of researchers (Bethel et al, 1984) using a colder temperature (-10°C) did find an additive response. Results of these two studies are summarized in Table 9-3.

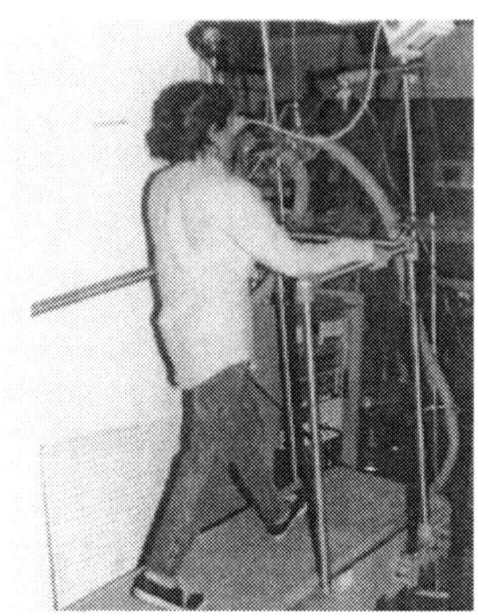

*Figure 9-1. An adolescent subject walking on a treadmill while breathing $SO_2$ in a controlled laboratory study.*

*Table 9-3. Interaction between cold, dry air and inhaled $SO_2$ on pulmonary function.*

| *Study:* | *Linn et al.* | *Bethel et al.* |
|---|---|---|
| *Subjects:* | *22 Mild asthmatics 18-31 years* | *7 Asthmatics, 24-36 years* |
| *$SO_2$ conc* | *0.6 ppm* | *0.5 ppm* |
| *Temperature* | *$5^o$ and $22^o$* | *$-10^o$ and $22^o$* |
| *Pulmonary Measures* | *Specific airway resistance (SRaw), $FEV_1$* | *SRaw* |
| *Results* | *No effect of $5^o$ temp* | *Significant increase in SRaw (222%)* |

**Duration of Exposure**

In the earlier studies, large changes in pulmonary function were seen after only 10 minutes of moderate exercise breathing the $SO_2$ atmosphere. Two contrasting effects of duration with $SO_2$ exposure have been documented. Short durations are sufficient to produce a response and longer durations do not produced greater effects. One study showed that as little as two minutes of $SO_2$ inhalation (1 ppm) during exercise caused significant bronchoconstriction, as measured by airway resistance. Another study showed that the increase in airway resistance after 10 minutes of exposure to 1 ppm $SO_2$ during exercise was not significantly increased when the exposure was extended to 30 minutes.

**Route of Exposure**

$SO_2$ is a highly water soluble gas and is rapidly taken up in the nasal passages during normal, quiet breathing. Studies in human volunteers found that, after inhalation of an average of 16 ppm $SO_2$, less than 1% of the gas could be detected at the oropharynx (Speizer and Frank, 1966). A number of more recent studies have shown that the degree of $SO_2$-induced bronchoconstriction is less after nasal inhalation than after oral inhalation (Kirkpatrick et al, 1982; Bethel et al., 1983; Linn et al, 1983; Koenig et al, 1985). This is not surprising. However it is surprising that inhalation of $SO_2$ causes such a dramatic bronchoconstrictor effect when it appears little of the gas actually reaches the bronchial airways. On the other hand, nasal uptake of $SO_2$ is not without adverse consequences for the upper respiratory system. Koenig and co-workers (1985) reported significant increases in the nasal work of breathing (measured by posterior rhinomanometry) in adolescent subject with asthma. A later study replicated this effect and found that an anti-histamine, chlorpheniramine, mitigated the nasal obstruction, as shown in Figure 9-2.

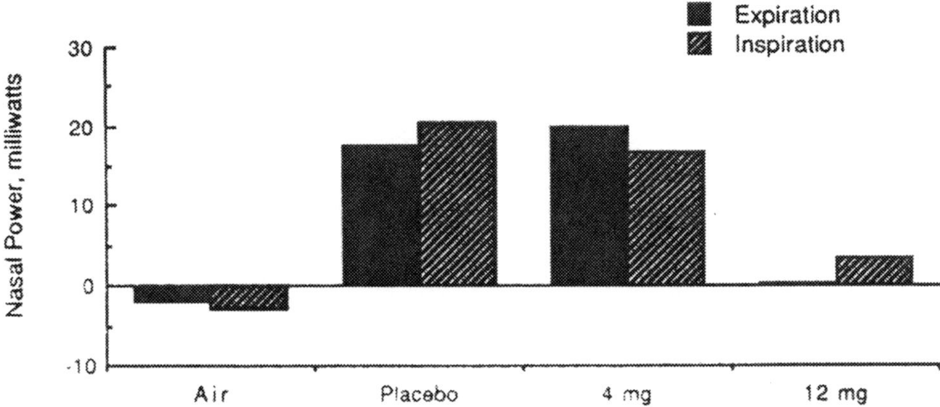


*Figure 9-2. Changes in nasal power after air or $SO_2$ exposure comparing chlorpheniramine with placebo treatments. FROM: Koenig et al, 1988b with permission.*

**Inflammation**

Sandstrom and co-workers (1989) reported inflammatory effects of $SO_2$ inhalation by evaluating bronchoalveolar lavage (BAL) fluid in healthy subjects. Both mast cells and monocytes were significantly elevated in BAL 4 and 24 hours after exposure to 8 ppm $SO_2$ for 20 minutes compared to air exposure. The mast cells showed a biphasic response with elevated numbers at 4 and 24 hours but not at 8 hours post exposure. The monocytes showed a lesser but continuous elevation. Also, Koenig and co-workers (1990) have shown, in a study of pulmonary function, that prior exposure to a sub-threshold concentration of ozone (0.12 ppm) potentiates the response to a subsequent exposure to low concentrations of $SO_2$ (0.1 ppm), a concentration that causes no pulmonary function decrements in single exposures. This result suggests that the ozone exposure altered bronchial hyperresponsiveness even though it did not alter pulmonary function. Whether the hyperresponsiveness was due to inflammatory changes was not assessed.

**Pharmacological Interventions**

The relationship between various therapeutic agents and $SO_2$ induced pulmonary function changes has been reviewed (Koenig and Pierson, 1991). Table 9-4 summarizes the results discussed in that review, updated to 1998.

*Table 9-4. Pharmacologic- $SO_2$ interactions*

| *Medication* | *Effect on $SO_2$ response* |
|---|---|
| *Albuterol* | *Completely blocks $SO_2$-induced bronchoconstriction Koenig et al., 1987; Linn et al, 1988)* |
| *Cromolyn Sodium* | *Blocks $SO_2$-induced bronchoconstriction at 60 mg Myers et al, 1986; Koenig et al, 1988a* |
| *Theophylline* | *Did not block $SO_2$-induced bronchoconstriction in patients on chronic theophylline.<br>Did mitigate $SO_2$-induced bronchoconstriction in a crossover study of Uniphyl (Koenig et al, 1992)* |
| *Chlorpheniramine* | *Did not block $SO_2$-induced bronchoconstriction. Did abolish the $SO_2$-induced increase in nasal work of breathing. (Koenig et al, 1988b).* |
| *Ipratropium Bromide* | *Did not block $SO_2$-induced bronchoconstriction; may have* |

| | |
|---|---|
| | *produced a slight mitigation (McManus et al, 1989)* |
| *Beclomethasone Dipropionate* | *No consistent effect on the* $SO_2$ *response. (Wiebicke et al, 1990)* |
| *Nedocromil Sodium* | *Inhibited* $SO_2$ *bronchoconstriction (Altounyan et al, 1986; Dixon et al, 1987; Bigby and Boushey, 1993).* |
| *Salmeterol* | *Statistically significant reduction of* $SO_2$*-induced decrement in* $FEV_1$ *from an average of 25 % to 7%. (Gong et al, 1996)* |

In spite of all the research investigating the relationship between $SO_2$ exposure and responses in individuals with asthma, the mechanism of the $SO_2$ response is not known. At one time it appeared, from animal studies, that $SO_2$-induced bronchoconstriction was mediated by the vagus nerve (part of the parasympathetic branch of the autonomic nervous system). Cooling or cutting the vagus nerve in cats abolished the $SO_2$ response. However as can be seen from Table 9-4, several therapeutic agents with varying sites of action inhibit the $SO_2$ response in human subjects. Also atropine, which counteracts the effects of the parasympathetic nervous system, does not inhibit the $SO_2$ response in human subjects. Thus there is not a clear understanding of why $SO_2$ elicits such a dramatic effect on the bronchial airways of subjects with asthma.

## Controlled Studies: Summary

In summary, controlled studies of human subjects with exposure to $SO_2$ have shown acute bronchoconstriction measured as decrements in $FEV_1$ or increases in airway resistance. Subjects with asthma show significant changes in pulmonary function at concentrations as low as 0.25 ppm $SO_2$ and after exposures as short as two and one-half minutes. On the other hand, subjects without asthma often show no significant change in pulmonary function after exposures in the range of 5.0 to 13 ppm $SO_2$ for approximately 30 minutes. Even though subjects with asthma show this dramatic response to inhaled $SO_2$, there is a tremendous inter-individual variation in responsiveness. In our laboratory we have seen marked differences in the pulmonary response to $SO_2$ in subjects with asthma in childhood, adolescence, young adulthood, and in seniors. In one study, the responses of adult subjects with asthma to a 10 minute $SO_2$ challenge test during a screening protocol demonstrated a wide variability in $FEV_1$ response to 0.5 ppm $SO_2$ (Trenga, 1997). The criterion for $SO_2$ sensitivity was at least an 8% decrease in baseline $FEV_1$ values immediately after the test. Forty-seven subjects completed the screening procedure. The change in $FEV_1$ after the $SO_2$ challenge ranged from a small increase (+6%) to a large decrement (-44%). Of the 47 subjects screened, 53% had a drop in $FEV_1$ greater or equal to 8%. Among those 24 subjects, the mean drop in $FEV_1$ was 17.2%. Total post challenge lower respiratory symptoms (cough, chest pain, wheeze, and shortness of breath) were significantly correlated with changes in $FEV_1$ ($p < 0.05$); FVC ($p < 0.05$); and PEF ($p < 0.01$). Total upper respiratory symptoms (sore throat and nasal discharge) after $SO_2$ challenges were not correlated with changes in pulmonary function. Also total post challenge lower respiratory symptoms in subjects who responded to $SO_2$ ($FEV_1$ drop greater or equal to 8%) were significantly greater ($p < 0.01$) than in the non-responding subjects. Decrements in $FEV_1$ were significantly greater ($p < 0.05$) in subjects who used both bronchodilators and anti-inflammatory medication, compared to those who just used daily bronchodilators.

There also is one study that offers data on the prevalence of $SO_2$ sensitivity in the general population. That recent study determined the prevalence of airway hyperresponsiveness to $SO_2$ in an adult population of 790 subjects aged 20-44 years, as part of the European Community Respiratory Health Survey (Nowak et al, 1997). The prevalence of $SO_2$ hyperresponsiveness (measured as a 20% decrease in $FEV_1$) in that population was 3.4%. The study did not attempt to diagnose subjects for the presence of asthma. Twenty two percent of subjects with a methacholine positive response showed $SO_2$ sensitivity while only 2 out of 679 who were not methacholine positive had such sensitivity. However another study did not find a correlation between $SO_2$ sensitivity and response to inhaled histamine (Magnussen et al, 1990).

## EPIDEMIOLOGIC STUDIES

Epidemiologic studies evaluating the effects of $SO_2$ on morbidity and mortality are often confounded by the high correlation between $SO_2$ and PM air pollution. In the eastern US most $SO_2$ is emitted from coal-fired electric power plants which also emit PM. The problems that arise due to this common source of these two pollutants and their high correlation in outdoor air regarding epidemiologic attribution of effects are discussed in Chapter 11. There is evidence concerning chronic exposure to $SO_2$ and adverse respiratory effects. Many Japanese cities, including Yokkaichi, experienced exposure to relatively high concentrations of $SO_2$ during the post WWII years as Japan was becoming a highly industrialized country. Several studies linked the $SO_2$ exposure with the increasing prevalence of respiratory disease. Although it is very difficult to use epidemiology to identify causation, in 1971 the Japanese courts accepted epidemiologic evidence showing a relationship between $SO_2$ and the prevalence of respiratory disease as legal proof of causation. Thus documentation of $SO_2$ exposure was sufficient for individuals to seek compensation under the Japanese Compensation Law (Namekata, 1986). This may be the only case where individuals are compensated for health effects caused by ambient air pollution. Figure 9-3 shows the relationship been $SO_2$ levels and mortality from asthma or chronic bronchitis from 1964 through 1982 (Dockery and Speizer, 1989). Environmental controls were placed on industry around 1971.

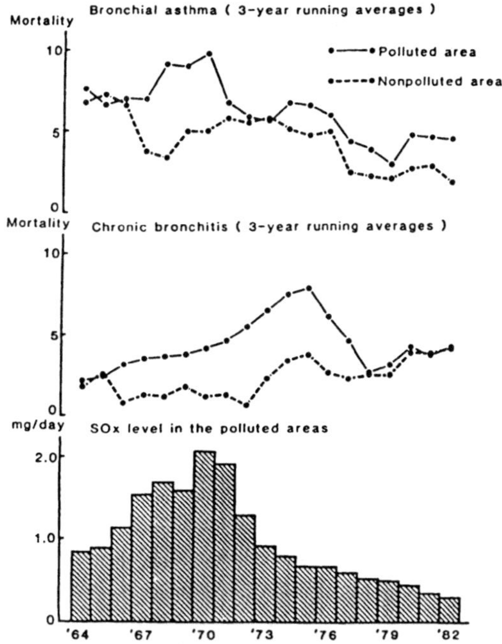


*Figure 9-3. Comparison of $SO_2$ levels and age-adjusted mortality due to asthma and chronic bronchitis in Yokkaichi, Japan. FROM: Dockery and Speizer, 1989 with permission.*

## SULFURIC ACID: CONTROLLED HUMAN STUDIES

Since $SO_2$ is a water soluble and reactive gas, it does not remain long in the atmosphere as a gas. Much of the $SO_2$ emitted is transformed through oxidation into acid aerosols, either sulfuric acid ($H_2SO_4$) or partially neutralized $H_2SO_4$ [ammonium bisulfate or ammonium sulfate] (see Figure 4-3). The ecological effects of acid aerosols (in the form of acid rain or dry deposition) have received much attention but are not the subject of this book. More recently there is growing concern about the human health effects of these aerosols (Lipfert et al, 1989; Spengler et al, 1990). One concern regarding sulfuric acid has been for subjects with asthma who are extraordinarily sensitive to $SO_2$. Early controlled exposure work showed that, in fact, adolescents with asthma were affected by inhalation of $H_2SO_4$ (Koenig et al, 1983). In that study, ten adolescent subjects with allergic asthma inhaled either 100 $\mu g/m^3$ $H_2SO_4$ or 100 $\mu g/m^3$ NaCl during 30 minutes of exposure at rest plus 10 minutes during moderate exercise. $H_2SO_4$ exposure was associated with significant decrements in $FEV_1$ (-8%) and maximal flow (-29%). More recent reports found significant pulmonary function decrements in $FEV_1$ and FVC after a lower concentration of $H_2SO_4$ (70 $\mu g/m^3$) (Hanley et al, 1992). Other studies have investigated healthy subjects and found no effects below 350 $\mu g/m^3$ (Utell et al, 1985). Thus it appears that subjects with asthma are unusually sensitive to inhaled $H_2SO_4$ as they are to inhaled $SO_2$.

One interesting finding regarding controlled human exposure to sulfuric acid is that the acid-induced pulmonary effect appears to remit with continued exposure. Koenig et al (1992) found that exposure to 35 or 70 $\mu g/m^3$ sulfuric acid for 45 minutes was associated with a significant decrement in pulmonary function in adolescents with asthma but that 90 minutes exposures to the same concentrations in the same subjects were not associated with a similar decrement. Morrow and coworkers (1994) also found that the sulfuric acid-induced decrement in both FVC and specific airway conductance was greater after 35 minutes of exposure than after 60 minutes. The reason for this apparent attenuation of effect is not known.

Controlled studies of sulfuric acid exposures in human subjects have shown effects on mucociliary clearance as well as the described changes on pulmonary function. One study carried out one-hour exposures of 100 and 1000 $\mu g/m^3$ concentrations of sulfuric acid in healthy, nonsmoking subjects (Leikauf et al, 1984). After exposure, the subjects inhaled radiolabelled iron oxide particles for 1-3 minutes. Clearance was measured by counting the number of particles retained in the lung. Although the sulfuric acid exposures did not affect pulmonary function measurement, exposures to either 1000 or 1000 $\mu g/m^3$ sulfuric acid did significantly alter bronchial mucociliary clearance of the 4.2 Mm iron oxide particles. Data from this study are shown in Figure 9-4 that shows that the actual calculated exposures were slightly different than the target concentrations (108 and 983 $\mu g/m^3$ rather than 100 and 1000).

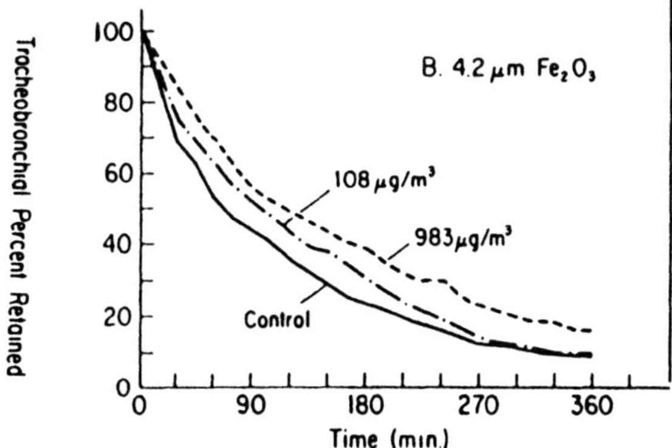


*Figure 9-4. Effect of sulfuric acid exposure on average bronchial mucociliary clearance in 10 healthy subjects. FROM: Leikauf et al, 1984 with permission.*

These data can be compared to a large literature on the effects of sulfuric acid on mucociliary clearance in rabbits. Those experiments also show that sulfuric acid exposure is associated with a significant slowing of clearance. The rabbit studies also show that clearance is depressed for up to six months after the end of the acid exposures. The implications of these findings are fairly important. Mucociliary clearance is one of the most important defense mechanisms of the lung as described in Ch 2. Sulfuric acid exposure also has been shown to affect the integrity of alveolar macrophages. Macrophages were retrieved from subjects exposed to 1000 $\mu g/m^3$ sulfuric acid for 2 hours. The cells were given a battery of tests to determine sulfuric acid-induced injury. Although the majority of the assays showed no difference between control and acid exposure, the alveolar macrophages did exhibit a significant depression of their normal oxygen-dependent killing mechanisms (Zelikoff et al, 1997). It should be noted that the exposure was a one-time event and thus not representative of human populations exposed of sulfuric acid aerosol throughout an entire summer.

## EPIDEMIOLOGY

A study of symptoms and medication use obtained from daily diaries completed by a panel of 207 adults with asthma was conducted in Denver (Ostro et al, 1991). Airborne hydrogen ion (a surrogate for sulfuric acid) was found to be significantly associated with both cough and shortness of breath in these subjects. A recent study on effects of acidic particles (primarily sulfuric acid) at a summer camp in the Austrian Alps found that the particles were associated with transient decreases in lung function in the children attending the camp Studnicka et al, 1995). Thus this study of true outdoor exposures agrees with the controlled laboratory studies described above.

More traditional epidemiologic studies also have implicated sulfuric acid (or at least some form of sulfate) as a suggestive cause of mortality. The Harvard Six City Study has provided the air pollution field with numerous analyses of adverse health effects. One fairly recent study using the Six City data found an association between air pollution and mortality. The Six Cities were ranked in terms of air pollution: the most polluted city of the six was Steubenville, Ohio and the least polluted was Portage Bay, WI. The other four cities were (from most polluted--Harriman, TN, St Louis, MO, Watertown, MA, and Topeka, KA). The analysis was a count of the risk of death in each city associated with PM air pollution, including sulfate air pollution. A total of 8111 adults were studied over a 14-16 year period. A strong relationship was seen between PM and mortality. The increased risk of death for subjects living in Steubenville was 26% greater than for those living in Portage Bay. However the association between mortality and sulfate was essentially the same as that of PM. Figure 9-5 shows the slopes of the relationships between mortality and both PM and sulfate. Judging from this study, one might conclude that sulfate is the toxic component of PM and thus responsible for the vast number of studies relating adverse cardiopulmonary effects and PM. However effects of PM have been seen in other regions of the US, namely Salt Lake City and Santa Clara County, CA where there are very low concentrations of sulfate. This fact has made many scientists skeptical that sulfate is, in fact, solely responsible. The controversy regarding the most toxic components of PM and the mechanisms of their action is covered in more detail in Chapter 11.

As with $SO_2$, mechanisms of action by sulfuric acid on the respiratory system are not known. Also of interest is the fact that sulfuric acid as an occupational exposure is categorized as a potential carcinogen. Sulfuric acid in the workplace is present at much higher concentrations than in ambient air. Also it is frequently in the form of a mixture with other pollutants. Without doubt, the complex mixture of air pollution containing $SO_2$, sulfuric acid and other PM is associated with adverse effects. What ability investigators will ultimately have in differentiating the effects of these pollutants individually is unknown.

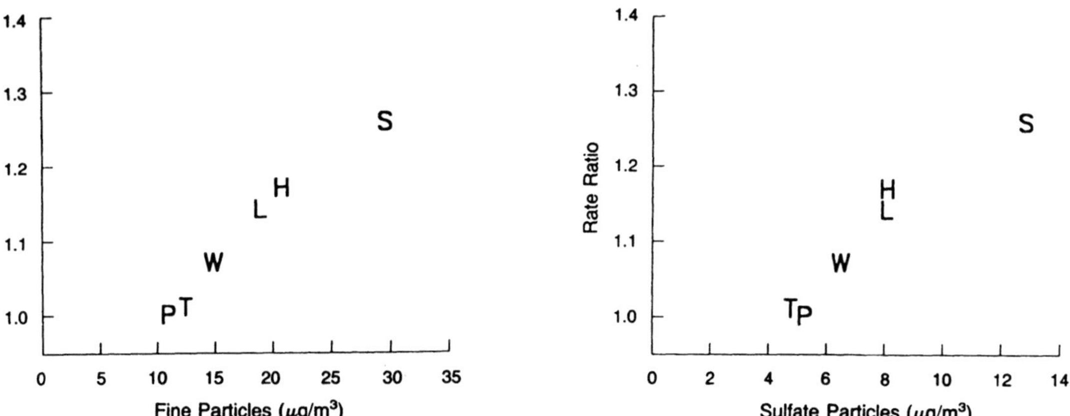


*Figure 9-5. Estimated adjusted mortality-rate ratios and pollution levels in the Harvard Six City study. FROM: Dockery et al, 1993 with permission.*

# CONCLUSION

In conclusion, regarding the respiratory effects of inhaled $SO_2$, the present NAAQS for $SO_2$ averaged over 24 hours does not seem sufficiently protective for the following reasons. 1) Bronchoconstrictive effects have been seen in asthmatic subjects after only a 2- minute exposure. 2) BAL studies and pollutant mixture studies (ozone followed by $SO_2$) have shown that, as well as $SO_2$-induced pulmonary mechanical responses, there are inflammatory responses in the bronchial airways. 3) Pharmocologic intervention studies raise concerns that common therapeutic doses of these agents can prevent $SO_2$-induced adverse respiratory effects. As stated in Chapter 4, EPA has set two health based standards for concentrations of $SO_2$ believed to be low enough to protect the public health; 0.14 parts per million (ppm) averaged over a 24 hour period and 0.03 ppm as an annual average. However, several states also have set shorter term standards for $SO_2$ based on air chemistry and meteorological data that show that down-drafts from $SO_2$ emitting stacks can increase ground level concentrations of this gas briefly (minutes to hours) which, when averaged over 24 hours, would not violate the 24 hour standard. For instance, Washington State has set the following short term standards for $SO_2$

*Table 9-5. Washington State ambient air quality standards for sulfur dioxide.*

| | | |
|---|---|---|
| *1 hour* | *0.25 ppm* | *Not to be exceeded more than twice in seven days* |
| *1 hour* | *0.40 ppm* | *Never to be exceeded* |
| *5 minute* | *1.00 ppm* | *Not to be exceeded more than once in eight hours.* |

*FROM: 1991 Air Quality Data Summary, Puget Sound Air Pollution Control Agency, 1991.*

# REFERENCES

Altounyan REC, Cole M, Lee 5:07 PM. Inhibition of sulphur dioxide-induced bronchoconstriction by nedocromil sodium and sodium cromoglycate in non-asthmatic, atopic subjects. Eur J Respir Dis 1986; 69 (suppl 147): 272-276.

Bethel RA, Erle DJ, Epstein J, et al. Effect of exercise rate and route of inhalation on sulfur-dioxide-induced bronchoconstriction in asthmatic subjects. Am Rev Respir Dis 1993; 128: 592-596.

Bethel RA, Sheppard D, Epstein J. Interaction of sulfur dioxide and dry cold air in causing bronchoconstriction in asthmatic subjects. J Appl Physiol 1984; 57: 491-423.

Bigby B, Boushey H. Effects of nedocromil sodium on the bronchomotor response to sulfur dioxide in asthmatic patients J Allergy Clin Immunol 1993; 92: 195-197.

Dixon CM, Fuller RW, Barnes PJ. Effect of nedocromil sodium on sulphur dioxide induced bronchoconstriction. Thorax 1987; 42: 462-465.

Dockery DW, Speizer FE. Epidemiological evidence for aggravation and promotion of COPD by acid air pollution. Lung Biol Health Dis 1989; 43: 201-225.

Dockery DW, Pope CA III, Xu X, et al. An association between air pollution and mortality in six US cities. New Engl J Med 1993; 329: 1753-1759.

Gong H, Linn WS, Shamoo DA, et al. Effect of inhaled salmeterol on sulfur dioxide-induced bronchoconstriction in asthmatic subjects. Chest 1996; 110: 1229-1235.

Hanley QS, Koenig JQ, Larson TV, et al. Response of young asthmatic patients to inhaled sulfuric acid. Am Rev Respir Dis 1992; 145: 326-331.

Kirkpatrick MB, Sheppard D, Nadel JA, Boushey HA. Effect of the oronasal breathing route on sulfur dioxide-induced bronchoconstriction in exercising asthmatic subjects. Am Rev Respir Dis 1982; 125: 627-631.

Koenig JQ, Pierson WE, Frank R. Acute effects of inhaled $SO_2$ plus NaCl droplet aerosol on pulmonary function in asthmatic adolescents. Environ Res 1980; 22: 145-153.

Koenig JQ, Pierson WE, Horike M, Frank R. Effects of $SO_2$ plus NaCl aerosols combined with moderate exercise on pulmonary function in asthmatic adolescents. Environ Res 1981; 25: 340-348.

Koenig JQ, Pierson WE, Horike M. The effects of inhaled sulfuric acid on pulmonary function in adolescent asthmatics. Am Rev Respir Dis 1983; 128: 221-225.

Koenig JQ, Morgan MS, Horike M, Pierson WE. The effects of sulfur oxides on nasal and lung function in adolescents with extrinsic asthma. J Allergy Clin Immunol 1985; 76: 813-819.

Koenig JQ, Marshall SG, Horike M, et al. The effects of albuterol on $SO_2$-induced bronchoconstriction in allergic adolescents. J Allergy Clin Immunol 1987; 79: 54-58.

Koenig JQ, Marshall SG, van Belle G, et al. Therapeutic range cromolyn dose-response inhibition and complete obliteration of $SO_2$-induced bronchoconstriction in atopic adolescents. J Allergy Clin Immunol 1988a; 81 part 1: 897-901.

Koenig JQ, McManus MS, Bierman CW, et al. Chlorpheniramine-sulfur dioxide interactions on lung and nasal function in allergic adolescents. Pediat Asthma Allergy Immunol 1988b; 2: 199-205.

Koenig JQ, Covert DS, Hanley QS, et al. Prior exposure to ozone potentiates subsequent response to sulfur dioxide in adolescent asthmatic subjects. Am Rev Respir Dis 1990; 141: 377-380.

Koenig JQ, Pierson WE. Air pollutants and the respiratory system: Toxicity and pharmacologic interventions. J Toxicol; Clin Toxicol. 1991; 29: 401-411.

Koenig JQ, Dumler K, Rebolledo V, et al. Theophylline mitigates the bronchoconstrictor effects of sulfur dioxide in subjects with asthma. J Allergy Clin Immunol 1992; 89: 789-794.

Koenig JQ, Covert DS, Larson TV, Pierson WE. The effect of duration of exposure on sulfuric acid-induced pulmonary function changes in asthmatic adolescent subjects: A dose-response study. Toxicol Indust Health 1992; 8: 285-296.

Leikauf GD, Spektor DM, Albert RE, Lippmann M. Dose-dependent effects of submicrometer sulfuric acid aerosol on particle clearance from ciliated human lung airways. Am Ind Hyg Assoc J 1984; 45: 285-292.

Linn WS, Shamoo DA, Spier CE,et al. Respiratory effects of 0.75 ppm sulfur dioxide in exercising asthmatics: influence of upper-respiratory defenses. Environ Res 1983; 30: 340-348.

Linn WS, Shamoo DA, Venet TG, et al. Comparative effects of sulfur dioxide exposures at $5^{o}$ C and $25^{o}$C in exercising asthmatics. Am Rev Respir Dis 1984; 129: 234-239.

Linn WS, Avol EL, Shamoo DA, et al. Effect of metaproterenol sulfate on mild asthmatics response to sulfur dioxide exposure and exercise. Arch Environ Health 1988; 43:399-406.

Lipfert FW, Morris SC, Wyzga RE. Acid aerosols: The next criteria pollutant? Environ Sci Technol 1989; 23: 1316-1322.

Magnussen H, Jorres R, Wagner HM, et al. Relationship between the airway response to inhaled sulfur dioxide, isocapnic hyperventilation, and histamine in asthmatic subjects. Int Arch Occup Environ Health 1990; 62: 485-491.

McManus MS, Koenig JQ, Altman LC, Pierson WE. Pulmonary effects of sulfur dioxide exposure and ipratropium bromide pretreatment in adults with nonallergic asthma. J Allergy Clin Immunol 1989; 83:619-626.

Morrow PE, Utell MJ, Bauer MA, et al. Effects of near ambient levels of sulphuric acid aerosol on lung function in exercising subjects with asthma and chronic obstructive pulmonary disease. Ann Occup Hyg 1994; 38 (suppl 1): 933-938.

Myers DJ, Bigby BG, Boushey H. The inhibition of sulfur dioxide-induced bronchoconstriction in asthmatic subjects by cromolyn is dose dependent. Am Rev Respir Dis 1986; 133: 1150-1153.

Namekata T. The Japanese compensation system for respiratory disease patients: background, status and trend, and review of epidemiological basis. IN Lee, Schneider, Grant and Verkerk (Eds) Aerosols Lewis Publishers, Chelsea MI. 1986.

Nowak D, Jorres R, Berger J, Claussen M, Magnussen H. Airway responsiveness to sulfur dioxide in an adult population sample. Am J Respir Crit Care Med 1997; 156: 1151-1156.

Ostro BD, Lipsett MJ, Wiener MB, et al. Asthmatic responses to airborne acid aerosols. Am J Public Health 1991; 81: 694-702.

Sandstrom T, Stjernberg N, Andersson M-C, et al. Cell response in bronchoalveolar lavage fluid after exposure to sulfur dioxide: A time-response study. Am Rev Respir Dis 1989; 140: 1828-1831.

Sheppard D, Wong WS, Uehara CF, et al. Lower threshold and greater bronchomotor responsiveness of asthmatic subjects to sulfur dioxide. Am Rev Respir Dis 1980; 122; 873-878.

Speizer FE, Frank R. The uptake and release of $SO_2$ by the human nose. Arch Environ Health 1966; 12: 725-728.

Spengler JD, Brauer M, Koutrakis P. Acid air and health. Environ Sci Technol 1990; 24: 946-956.

Studnicka MJ, Frischer T, Meinert R, et al. Acidic particles and lung function in children. Am J Respir Crit Care Med 1995; 151; 423-450.

Trenga CA. Dietary antioxidants and ozone-induced bronchial hyperresponsiveness in adult asthmatics. PhD Thesis, University of Washington, Department of Environmental Health, Seattle, 1997.

Utell MJ. Effects of inhaled acid aerosols on lung mechanics: An analysis of human exposure studies. Environ Health Perspect 1985; 63: 39-44.

Wiebicke W, Jorres R, Magnussen H. Comparison of the effects of inhaled corticosteroids on the airway response to histamine, methacholine, hyperventilation, and sulfur dioxide in subjects with asthma. J Allergy Clin Immunol 1990; 86: 915-923.

Zelikoff JT, Frampton MW, Cohen MD, et al. Effects of inhaled sulfuric acid aerosols on pulmonary immunocompetence; A comparative study in humans and animals. Inhal Toxicol 1997; 9: 731-752.

# CHAPTER 10.

# HEALTH EFFECTS OF PARTICULATE MATTER

Particulate matter (PM) air pollution is the most complex pollutant of the six pollutants regulated by the US EPA as "criteria" air pollutants. Particulate matter is all the dust, smoke, and haze particles suspended in ambient air as described in Chapter 4. The reader is referred back to Figures 4-4a and 4-4b to review the distributions of atmospheric particles by size. Chapter 4 also lists common sources of PM. Briefly, as described in chapter 4, there are basically two types of sources of PM: a so called "coarse fraction" and a finer fraction. The coarse fraction is made up of natural aerosolization of crustal matter such as during a wind blown dust storm, during agricultural practices, and during excavations for construction, whereas the source of the finer fraction is mainly combustion including gasoline and diesel fuel vehicle combustion, industry combustion, coal combustion in the process of electricity generation, and burning of vegetative material such as wood burning for residential heating and grass burning for clearing agricultural land. The size of PM also is expressed as the difference between the coarse and the fine mode ($PM_{10-2.5}$ or $PM_{CF}$). The smallest particles in air are the ultrafine particles that are emitted from combustion and are generally smaller than a few hundred nanograms.

## DEPOSITION

The health effects of PM are dependent on the interactions between PM and the respiratory system. As we inhale air containing various types and concentrations of PM some of this material is deposited in the nasal and bronchial airways. Deposition of particles in the nose and lungs is mainly dependent on the size of the particles. This size dependence is shown nicely in Figure 10-1.

There are four processes by which particles deposit in the nose or lungs. These are sedimentation, inertial impaction, diffusion, and static electrical forces. Sedimentation is a primary process by which larger particles land in the airways--they simply fall out of the air stream due to the size and weight (gravitational forces). Inertial impaction is a process by which particles slam against the walls of the airways at bifurcations as the particle laden air stream flows into the upcoming branch. Diffusion (or simple Brownian movement) is the process by which the smallest particles deposit in the airways. This is the primary process for deposition in the periphery of the lung. As inhaled air flow decreases in the furthest regions of the lungs and as the diameter of the airways narrows, there is more opportunity for small particles to diffuse to the airway walls. Electric charge is not a common process by which particles deposit in the lungs; however it does occur when particles in the airstream have an electric charge. The deposition processes have an influence on health effects. For instance, the primary sites of cancer in the bronchi are at points of bifurcation. It is believed this is due to the increased number of particles that impact at the bifurcations. Deposition patterns in the airways are just one of the many reasons why the diameter of particles in ambient air become as issue when air quality standards are being written. Particles that are hygroscopic, that is take-up water, grow in the airways as they are inhaled. This affects their diameter and the subsequent site of deposition. A good example of hygroscopic growth is sulfuric arid droplets, which can grow 2-3 times in size upon inhalation into the humid airways.

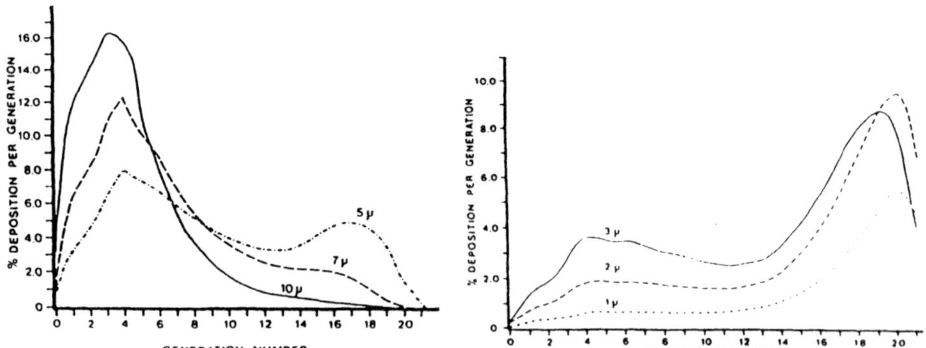


*Figure 10-1. Deposition of particles of different sizes as a function of airway generation. Generation number begins with 0 indicating the trachea and increasing to number 21 which represents the periphery of the lung, the alveolar spaces. FROM: Gerrity et al, 1979 with permission.*

## CLEARANCE

There are three primary processes involved in the clearance of particles from the lungs. First, the presence of mucus and ciliated epithelial cells lining the airways create a mucociliary escalator. Particles land on the mucus lining of the airway; the movement of the ciliated cells which line the airways propel the mucus lining mouthward thus removing those particles from the lung (see Figure 10-2). Secondly, particles that land in the periphery of the lungs are often engulfed by alveolar macrophages. Engulfment physically removes the particle from the airway. In addition, macrophages contain enzymes that can detoxify certain particles. Thirdly, cough is considered a lung defense mechanism. When we cough, considerable pressure is created which can expel particles from the lung. The time frame in which these defenses operate varies, with cough being instantaneous, mucociliary clearance requiring hours to days, and removal by macrophages taking from days to weeks.

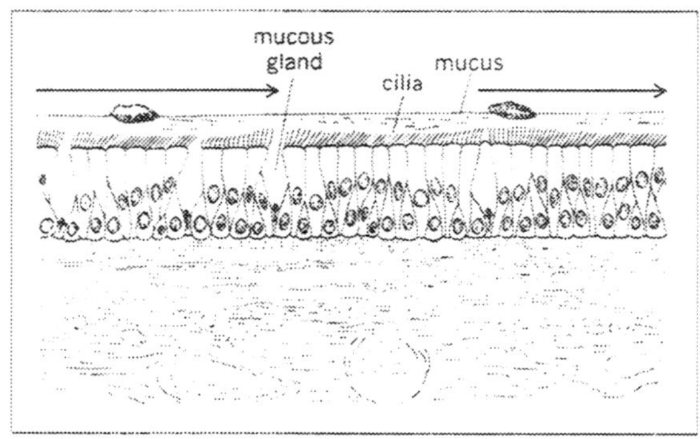


*Figure 10-2. Epithelial lining of the respiratory tract and the muco-ciliary escalator. The arrows indicate the upward direction in which the cilia move the overriding layer of mucus to which foreign particles are stuck. FROM: Vander et al, 1970 with permission.*

## PM NAAQS

In order to understand the progression of studies pertaining to the health effects of PM one needs to understand the various measures of PM that are or have been available for use in health effects studies. The original National Ambient Air Quality Standard (NAAQS) for PM set in the early 1970s was for total suspended particulate matter (TSP). In 1987, as a result of a better understanding of the size of particles respired into the lung, the US EPA revised the PM standard changing the size of the PM metric to equal or less than 10 μm in diameter. This metric is referred to as $PM_{10}$. Most recently, in July 1997 EPA added an additional measure of PM to the NAAQS, a measurement of even finer particles equal or less than to 2.5 μm in diameter. The concentrations and averaging times for these three PM metrics are shown in Table 10-1.

*Table 10-1. Various measures of PM air pollution.*

*TSP*
*260 μg/m³ for 24 hours*
*150 μg/m³ annually.*

*$PM_{10}$*
*150 μg/m³ for 24 hours*
*50 μg/m³ annually.*

*$PM_{2.5}$*
*65 μg/m³ for a 24 hr average*
*15 μg/m³ for an annual average*

---

There are several other measures of PM that a reader will encounter in the literature on health effects of PM. Three of these are coefficient of haze (COH), black smoke (BS), and light scattering coefficient ($\sigma_{sp}$).

## HEALTH EFFECTS OF PM

### Epidemiology

As mentioned earlier, adverse health effects can be categorized into morbidity (illness) or mortality (death). Mortality studies will be reviewed first. In the winter of 1952, a cold weather inversion in London trapped smoke from coal burning at the ground level. As the inversion continued, increasing numbers of deaths were noted. A view of the association between air pollution --particulate matter (measured as black smoke) and $SO_2$--and deaths is shown in Figure 10-3.

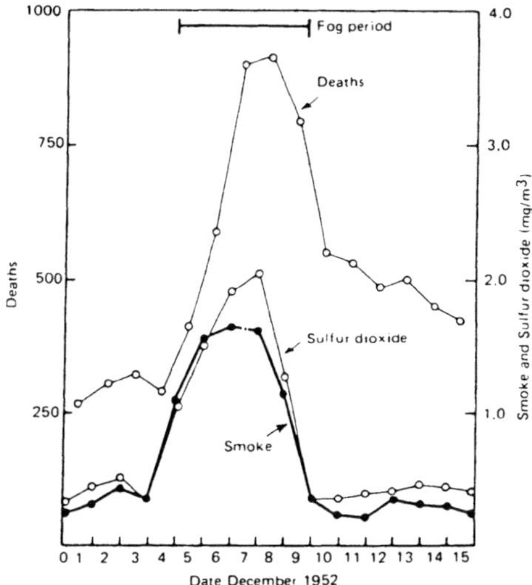


*Figure 10-3. Smoke concentrations, $SO_2$ concentrations and deaths in London in the winter of 1952. FROM: Dockery and Speizer, 1989 with permission.*

There are several relationships to note in this figure. First, note that the number of deaths increase after the air pollutants begin to increase. This satisfies the criteria of temporality put forth by Brandon Hill, as cited in Ch 7. Also, note that concentrations of the pollutants (on the right hand Y axis) are given in $mg/m^3$. Current levels of PM and $SO_2$ in urban areas are measured in $\mu g/m^3$, an order of magnitude lower that these values from London. Thirdly, note that the rate of deaths does not return to the pre-episode level by the end of the time period depicted, suggesting a lasting effect of the increased smoke levels. Records show that at least 4000 excess deaths occurred during the one-week period of heightened pollution. British citizens began to demand governmental action to decrease such air pollution. Notice of this episode and others was also taken in the US; during the 1950s and 1960s there were various attempts to write legislation to control air pollution. However, it was not until after the passage of the US Clean Air Act in 1970 and the setting of NAAQS for the most wide-spread air pollutants, that air pollutant levels began to decrease in the US. Figure 4-3 shows the decreases in air pollutant levels between 1970 and 1990. Due to data similar to these, it was generally considered that PM levels in the US in 1980s and 1990s were not high enough to have adverse health effects. However in 1991 Joel Schwartz and associates published papers showing that PM air pollution in Philadelphia from 1973-1980 and in Steubenville, Ohio from 1974-1984 was in fact associated with daily mortality. These studies followed the time series epidemiologic design described in Chapter 7. A graph showing the linear relationship between the relative risk of death and the concentration of PM is given in Figure 10-4. PM during that time period was measured as TSP as described earlier in this chapter. The conclusion of those studies was that daily mortality increased approximately 10% for each increase of 100 $\mu g/m^3$ in TSP. (This would translate to approximately a 50 $\mu g/m^3$ increase in $PM_{10}$.)

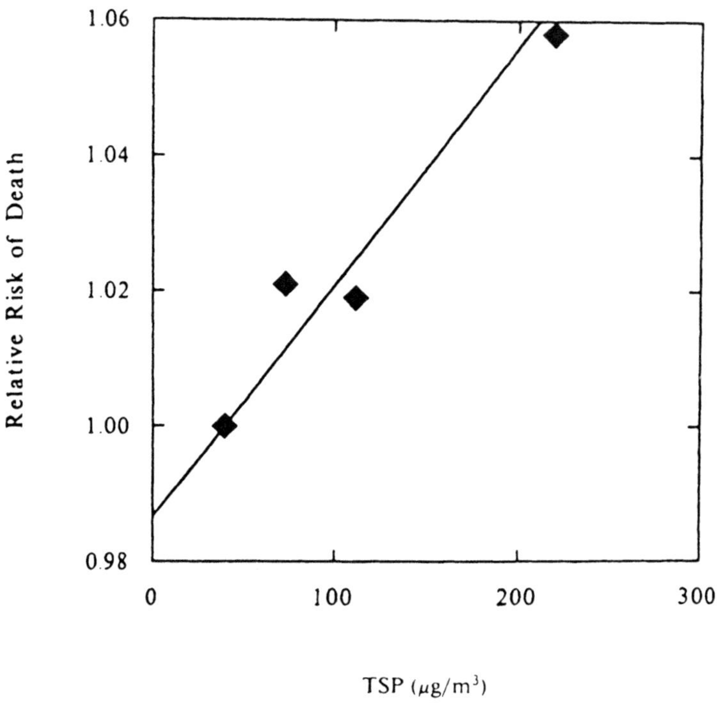


*Figure 10-4. Relative risk of death in Steubenville, Ohio by quartile of TSP. FROM: Schwartz and Dockery, 1992 with permission.*

When Joel Schwartz presented the data on Philadelphia at a scientific meeting, the New York Times published an article based on that research estimating that up to 60,000 excess deaths a year could occur in the US as a result of PM air pollution. This estimate came from applying the Philadelphia data to the national air pollution data. Joel Schwartz was quoted as saying " for people who are old or sick, particulate matter may be the straw that breaks the camels back". Many people were extremely critical of these estimates, however more studies were published and there was a remarkable consistency in the estimates of increased mortality from city to city. In 1994 Dockery and Pope published a review of the respiratory effects of PM. Table 10-2 is taken from that report and lists various health outcomes with the percent increase associated with a 10 $\mu g/m^3$ increase in $PM_{10}$.

*Table 10-2. Combined Effect Estimates of Daily Mean Particulate Pollution*

| | % Change in health indicator per each 10/μg/m³ increase in $PM_{10}$ |
|---|---|
| *Increase in daily mortality* | |
| *Total deaths* | *1.0* |
| *Respiratory deaths* | *3.4* |
| *Cardiovascular deaths* | *1.4* |
| *Increase in hospital usage (all respiratory)* | |
| *Admissions* | *0.8* |
| *Emergency department visits* | *1.0* |
| *Exacerbation of asthma* | |
| *Asthmatic attacks* | *3.0* |
| *Bronchodilator* | *2.9* |
| *Emergency department *visits* | *3.4* |
| *Hospital admissions* | *1.9* |
| *Increase in respiratory systems reports* | |
| *Lower respiratory* | *3.0* |
| *Upper respiratory* | *0.7* |
| *Cough* | *1.2* |
| *Decrease in lung function* | |
| *Forced expired volume* | *0.15* |
| *Peak expiratory flow* | *0.08* |

**one study only*
*FROM: Dockery and Pope, 1994 with permission.*

Bates has elegantly pointed out that when one is trying to decide whether there is an association between air pollutant and health outcomes, it is important to look for coherence in the data (1992). A glance at Table 10-2 should convince the reader that there is a coherence of health outcomes associated with PM. Not only is mortality associated with PM levels, but many indicators of disease aggravation prior to an actual death--increased hospital use, decreased lung function, and increased symptoms of distress also are associated with elevated PM concentrations. These are logical progressions from mild to moderate aggravation of disease all the way to a fatal illness. This coherence of effects is another of Brandon Hill's criteria for causation (Ch 7).

Two more recent epidemiologic studies of mortality and PM used the cohort design described in Chapter 7. The Harvard Six Cities studies have been collecting air pollution and health outcome data since the early 1970s. Dockery and co-workers (1993) used these data, which include over 8000 individuals, to test for survival in six cities ranked from most to least polluted: Steubenville, OH; St Louis, MO; Harriman, TN; Watertown, MA; Topeka, KN; and Portage, WI. In contrast to the time series studies described above, personal characteristics of the Six City subjects were known and cigarette smoking and other important variables were controlled. Figure 9-5 shows the probability of survival in each city over a 16 year time period associated with nearly identical slopes for both $PM_{2.5}$ and sulfate. The study found that there was a 26% greater chance of dying in Steubenville (the most polluted city) compared to Portage (the least polluted city). The measure of particulate matter in the study was $PM_{2.5}$. As mentioned, the study found that using a measure of sulfate alone led to very similar estimation of risk. However, other studies in cities without significant levels of sulfuric acid found similar risks of PM. The jury is still out on the relative contribution of sulfates to the PM effect. There are many other constituents of PM and each of these has not been tested against mortality. A tabular representation of the Six City data is shown in table 10-3. This table shows the average annual $PM_{2.5}$ levels and the percentage of increased risk for early death in each city.

*Table 10-3. Annual average $PM_{2.5}$ values compared to the percentage of increasing risk of early death when compared with the cleanest city, Portage.*

| *City* | *Average annual $PM_{2.5}$ $\mu g/m^3$* | *Increased risk of death %* |
|---|---|---|
| *Portage* | *11.0* | *0* |
| *Topeka* | *12.5* | *1* |
| *Watertown* | *14.9* | *7* |
| *St Louis* | *19.0* | *14* |
| *Harriman* | *29.6* | *26* |

EPA's decision to add a new fine particulate matter standard ($PM_{2.5}$) was based largely on the Six City Study. One other cohort study that also used $PM_{2.5}$ as a metric was a large study conducted in collaboration with the American Cancer Society (ACS). The ACS had available a large data set (552,138 individuals) recruited in the mid 1970s and followed for a number of years (Pope et al, 1995). Sulfate and fine particles were estimated from air agency data bases. The relationships of air pollution to all-cause, lung cancer, and cardiopulmonary mortality were examined using multivariate analyses that controlled for smoking, education, and other risk factors. Adjusted relative risk ratios and associated confidence intervals for all cause mortality for the most polluted areas compared to the least polluted areas were 1.15 (1.09 to 1.22) for sulfate and 1.17 (1.09 to 1.26) for fine particles. When specific causes were examined, particulate matter was associated with cardiopulmonary mortality and lung cancer.

Even with these two large studies there still is much debate about the relative effect of particle size on mortality (and other health outcomes). A fairly recent paper provides some data that can be applied to that debate. Schwartz and associates (1997) compared the effect of $PM_{2.5}$, and $PM_{10}$ and coarse matter (CM), the mass between 2.5 and 10 µ,m, as predictors of mortality using the six cities in the Six City study. Thus the study was designed to be a direct test of the relative risk of finer versus coarser particles. $PM_{2.5}$ showed the most significant association with mortality in this data set. Data from that study are shown in Table 10-4. Effect of the association with particle size was estimated separately and long-term trends were controlled.

*Table 10-4. Estimated increase in daily mortality, 95% CI, and t statistic by city and combined estimated associated with a 10 µg/m³ increase in PM.*

| Study City | $PM_{2.5}$ | CM | $PM_{10}$ |
|---|---|---|---|
| Boston | 2.2% (1.5%, 2.9%) t = 6.31 | 0.2% (-0.6%, 1.2%) t = 0.58 | 1.2% (0.7%, 1.7%) t = 4.86 |
| Knoxville | 1.4% (0.2%, 2.6%) t = 2.26 | 1.0% (-0.6%, 2.6%) t = 1.20 | 0.9% (0.1%, 1.8%) t = 2.21 |
| St. Louis | 1.1% (0.4%, 1.7%) t = 3.17 | 0.2% (-0.7%, 1.1%) t = 0.45 | 0.6% (0.1%, 1.0%) t = 2.42 |
| Steubenville | 1.0% (-0.1%, 2.1%) t = 1.79 | 2.4% ( 0.5%, 4.3%) t = 2.43 | 0.9% (0.1%, 1.6%) t = 2.17 |
| Portage | 1.2% (-0.3%, 2.8%) t = 1.64 | 0.5% (-1.2%, 2.3%) t = 0.57 | 0.7% (-0.4%, 1.7%) t = 1.22 |
| Topeka | 0.8% (-2.0%, 3.6%) t = 0.53 | -1.3% (-3.3%, 0.6%) t = -1.32 | -0.5% (-2%, 0.9%) t = -0.67 |
| Combined | 1.5% (1.1%, 1.9%) t = 7.41 | 0.4% (-0.1%, 1.0%) t = 1.48 | 0.8% (0.5%, 1.1%) t = 5.84 |

*FROM: Schwartz et al, 1996 with permission.*

As indicated in Table 10-2, many other adverse health endpoints as well as mortality have been associated with PM, such as pulmonary function decrements, respiratory symptoms, increases in medication use in subjects with asthma, emergency department visits, and hospital admissions. These indicators of morbidity are reviewed at this point.

## PM and Pulmonary Function

Several studies have shown that acute exposure to PM is associated with decrements in pulmonary function. One study evaluated pulmonary function in elementary school children over a 9 month period, before, during and after the wood smoke heating season in Seattle (Koenig et al, 1993). Pulmonary function was measured approximately once per month in 326 children. Pulmonary function was associated with the PM level measured from 7 pm to 7 am for the day preceding the measurements. In general there was a 1.8 ml decrement in $FEV_1$ and FVC for each µg/m³ $PM_{2.5}$. This association was seen in the 26 children with asthma but not in the children without asthma. This result can be compared to findings of a study in Steubenville that found a relationship between FVC and TSP in healthy children (Dockery et al, 1989), and also with a study in Salt Lake City that showed a relationship between peak flow decrements and increases in $PM_{10}$ in subjects with asthma (Pope et al, 1991). All of these studies lacked personal exposure assessments and relied on measurements from fixed site monitors. However, a recent study reports that the correlation between the fixed and personal monitors is fairly high if distance from a busy roadway and exposure to ETS during the past 24 hours are taken into account (Janssen et al, 1998). In order to judge better the actual exposure to PM in a group of subjects, Silverman (1992) fitted 36 subjects with asthma with portable multipollutant samplers. The samplers were worn for 10 days in both the summer and winter. Associations between PM and lung function ($FEV_1$, FVC, and $FEF_{25-75}$) were significant during the winter period ($p < 0.04$). Daily medication use also increased significantly in the winter, complicating the interpretation.

## PM and Respiratory Symptoms

Dockery and associates (1989) reported a significant relationship between $PM_{15}$ and chronic cough, bronchitis, and chest illness in a group of children studied in the Harvard Six City Study. Schwartz and co-workers (1991) reported increased incidence of bronchitis and croup reported by hospitals, clinics, and private physicians in five German cities. An increase from 10 $\mu g/m^3$ to 70 $\mu g/m^3$ TSP was associated with a 27% increase in croup. Significant associations were found between respiratory symptoms and airborne hydrogen ion, an estimate of acid aerosols, in a population of adult asthmatics in Denver (Ostro et al, 1991). In that study, a panel of 207 patients with asthma recorded respiratory symptoms and medication use in daily diaries in the winter of 1987-1988. Airborne $H^+$ was significantly associated with moderate and severe cough ($p < 0.01$) and shortness of breath ($p < 0.05$).

## PM and Medication Use

Pope and co-workers (1991) have examined various health endpoints in residents of the Salt Lake City area where PM is emitted primarily by a steel mill. The meteorology of that region results in accumulations of particles in the air when temperatures are cold and stagnant conditions exist. Increases in medication use were associated with $PM_{10}$ in Utah Valley, Utah in a panel of subjects with asthma. In that study, the probability of increased medication use was 6.2 times higher when the 24 hour $PM_{10}$ levels were at the highest compared to the medication use during a 24 hour period at the lowest $PM_{10}$ level. With respect to the earlier discussion of $SO_2$ and PM covariance, the Utah airshed had essentially no $SO_2$ and thus is a test of the independent effects of $PM_{10}$. Studies in Germany and Austria also have found significant relationships between medication use in subjects with asthma and daily PM levels (Peters et al, 1996).

## PM and Emergency Department Visits

Bates and associates (1990) reported a significant association between both $SO_2$ and sulfuric acid (a form a PM) in visits to emergency departments for asthma. A study in Seattle found a strong association between visits to emergency departments for asthma at eight hospitals and $PM_{10}$ from 1989-1990 (Schwartz et al, 1993). The relative risk for a 30 $\mu g/m^3$ $PM_{10}$ increase was 1.12 (95% confidence interval 1.20 to 1.04) ($p < 0.005$). Figure 10-5 shows the relative risk of an emergency room visit for asthma by quartile of $PM_{10}$. During the period of the study, the highest daily concentration of $PM_{10}$ was 103 $\mu g/m^3$ well below the National Ambient Air Quality Standard set by EPA (150 $\mu g/m^3$), suggesting that the present standard is not sufficiently protective of susceptible groups such as patients with asthma. Similar associations have been seen for asthma visits to emergency departments in Santa Clara County, CA, and Barcelona. Asthma epidemics in Barcelona make an interesting detective story (Anto et al, 1989). It was observed that there were days on which episodes of asthma visits to emergency departments reached epidemic levels. After much careful examination, it was determined that the asthma epidemics were correlated with soybean dust from unloading ships in the harbor. This conclusion was confirmed when, upon installation of dust filters at the silos, the asthma episodes disappeared.

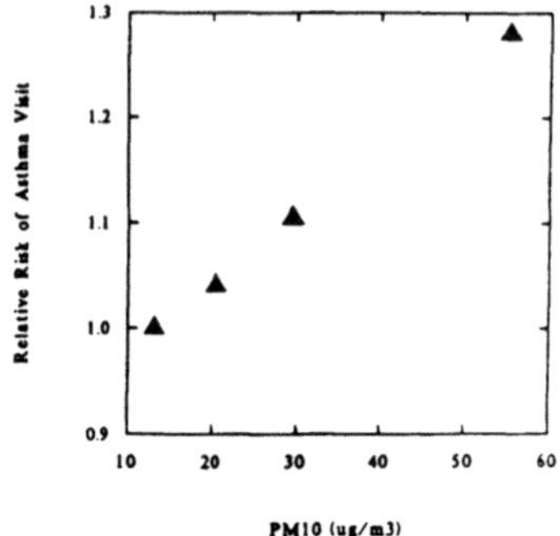


*Figure 10-5. Relative risk of asthma visits by quartile of $PM_{10}$ concentration plotted against the mean $PM_{10}$ concentration in the quartile. FROM: Schwartz et al, 1993 with permission.*

## PM and Hospital Admissions

Ponka (1991) tallied all hospital visits for asthma, both from emergency departments and admissions in Helsinki. When the pattern of the visits was compared to concentrations of outdoor air pollutants (nitric oxide, nitrogen dioxide, $SO_2$, carbon monoxide, ozone and TSP), nitric oxide and ozone levels were strongly associated with asthma but TSP concentrations were not. A recent study in Seattle tested the relationship between measured ambient air pollutants and non-elderly hospital admissions with a principal diagnosis of asthma (Sheppard et al 1999). The study found an estimated 4-5% increase in the rate of asthma hospital admissions associated with an interquartile range change in $PM_{10}$, $PM_{2.5}$ and $PM_{CF}$ with a lag period of one day between the pollution measurement and the hospital admission. The interquartile range is the average concentration of a pollutant in the range of 25-75% of the 24-hour average concentration. This study also found that CO was associated with the asthma admissions. Since there in no known relationship between CO and asthma aggravation, this finding suggests that CO was acting as a surrogate of air pollution in general. This result again emphasizes the difficulty in attributing health outcomes to just one of many air pollutants when their levels are highly correlated (Moolgavkar and Luebeck, 1996). Burnett and co-workers (1997) also found that both gaseous pollutants and various measures of PM were associated with hospital admissions for cardiorespiratory diseases in Toronto.

One of the most elegant studies of the relationships between hospital admissions and PM was conducted by chance. A steel mill in Utah was closed for a period one winter due to a strike among workers. Bronchitis and asthma hospital admissions were tallied for the winter of the strike and for the preceding and subsequent winters (Pope, 1991).

Both the PM concentrations and the hospital admissions were significantly lower during the winter of the strike. This natural experiment in controlled epidemiology gave very compelling results that are shown in Figure 10-6a and 6b.

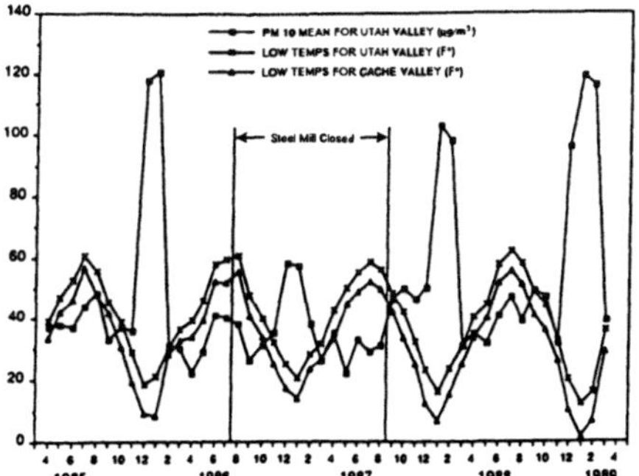


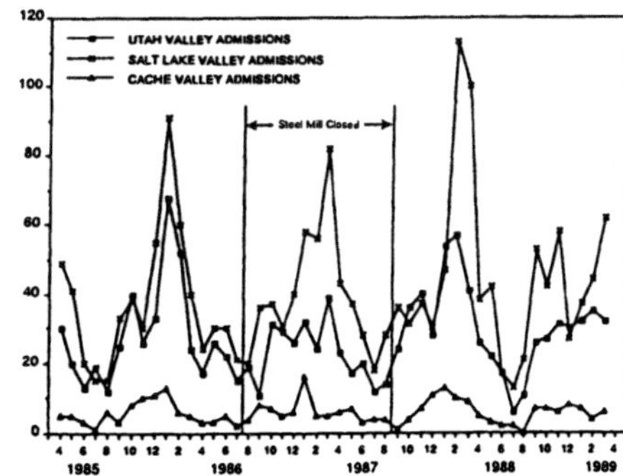


*Figure 10-6. a. Monthly mean $PM_{10}$ and temperature for Utah Valley, 1985 through 1989. b. Monthly mean hospital admissions for bronchitis and asthma for all ages in three Utah cities. FROM: Pope, 1991 with permission.*

## Wood Smoke

Particulate matter from wood smoke is a special category of very fine particles. The majority of particles generated from the incomplete combustion of wood in wood stoves is under 1.0 μm in diameter and thus easily inhaled deep into the lung. A summary of the emissions from wood smoke and their noncancer health effects was published recently (Larson and Koenig, 1994). Table 10-5 summarizes nine studies showing various health outcomes associated with exposure to residential wood smoke (see original article for citations). One epidemiologic time series study, published in 1990 using data from 1986 in Santa Clara County, California, reported an association between mortality and PM for winter months (Fairly, 1990). Much of the PM in the winter in this region comes from wood burning for residential heat (see Tables 4-7, 4-8, and 4-9). Indoor concentrations of wood smoke are much higher in some developing countries where wood in used for cooking food. A study investigating exposure to wood smoke and chronic airway disease in women in Mexico found odds ratios for wood smoke and chronic bronchitis equal to 3.9 and for wood smoke exposure and chronic bronchitis plus airway obstruction equal to 9.7 (Perez-Padilla et al, 1996).

Few of the studies cited in Table 10-5, had adequate measurements of wood smoke exposure. None of them used personal monitoring. Assessment of individual exposure to wood smoke (and other air pollutants) would be more accurate if a biomarker of exposure was available. Such a biomarker for wood smoke is under development as discussed in Ch 5 (Zhu, 1998). The compounds being explored as potential biomarkers for wood smoke exposure are a group of methoxylated phenols. Several methoxylated phenols are known to be pyrolysis products of lignin, particularly syringols and guaiacols. Lignin is the component of wood that imparts its strength. The biomarker can be detected in human urine at concentrations of wood smoke known to occur in neighborhoods. Thus this biomarker satisfies most of the criteria for a useful marker: sufficient sensitivity to detect acute exposures resulting from air pollution episodes; specific to wood smoke; and accurate and stable. It is expected that studies in the near future will validate the use of this biomarker for air pollution epidemiologic studies.

*Table 10-5. Summary of Studies of Respiratory Effects of Exposure to Wood Smoke.*

| Reference | Age (yr) | Number of subjects | Endpoints measured | Results |
|---|---|---|---|---|
| 41 | 1–7 yr | 34 w/stoves<br>34 without | Symptoms | More symptoms in children with stoves $p < 0.001$ |
| 93 | 5–11* | 258 w/stoves<br>141 without | Symptoms | Risk ratio = 1.1, showing no significant effect |
| 11 | 1 and older | 455 high smoke**<br>368 low smoke | Symptoms disease prevalance | No significant effects. Trend in children aged 1–5 |
| 54 | 8–11 | 296 healthy***<br>30 asthmatic | Spirometry | Significant association between fine particles and lung function in asthmatics, $p = 0.05$ |
| 14 | 1–5 ½ | 59 | Symptoms | Significant correlation between woodstove use and wheeze and cough frequency $p = 0.01$ |
| 68 | <24 mo | 58 pairs | Respiratory disease | Woodstove significant risk factor for lower resp infection |
| 36 | 8–11 | 410*** | Spirometry | Significant decrease in PFTs with elevated wood smoke |
| 45 | 8–11 | 495 | Spirometry | Significant relation of functional decrease with elevated wood smoke |
| 64 | $\bar{X} = 46$ | 182 | Symptoms | Significant association |
| 82 | all ages | 2955, asthma<br>3810, gastroenteritis | Emergency | Significant association between room visits asthma visits fine particles |

*Described as kindergarten through grade 6.
**Geographical areas with high or low wood smoke pollution.
***Grades 3 through 6.

FROM: Larson and Koenig, 1994 with permission

## Diesel

Diaz-Sanchez and co-workers have performed a series of experiments in human subjects examining the effects of diesel exhaust particles (DEP). Nasal lavage was the primary health assessment tool. Before the DEP challenge, nasal lavage fluid had detectable levels of only three mediators of inflammation. Eighteen hours after DEP exposure, in the form of a nasal spray, mRNA was detected for seven major cytokines known to be involved in inflammation (Diaz-Sanchez et al, 1997). Increases in IgE levels in lavage fluid also were seen. Another study found DEP bound to grass pollen particles which represents a possible mechanism by which allergens can become concentrated in polluted air and aggravate asthma (Knox et al, 1996). A review of the DEP literature found general evidence in animals and humans that DEPs enhance IgE production (Nel et al, 1998). Numerous animal studies (at very high concentrations) have seen increased lung tumors following diesel exhaust. The data are stronger for lung tumors in rats than in mice. Also five studies investigating tumor growth in hamsters exposed to diesel exhaust have shown negative results. Epidemiologic studies of cancer risk in human populations exposed to diesel exhaust suggest that long-term exposure is associated with a 20-50% increased risk for lung cancer. Due to evidence of this nature, the California Air Resources Board voted to identify diesel exhaust as a toxic air contaminant, paving the way for the state to establish more stringent regulations for state diesel exhaust limits (Leclair, 1997). This seems quite appropriate, since diesel engines release over 100 times more particles per mile traveled than do gasoline engines.

# PM AND CARDIOVASCULAR DISEASE

As seen in Table 10-2, PM adverse effects are not confined to aggravation of respiratory diseases. The time series studies of mortality and PM have identified cardiovascular disease as one of the group of diagnoses associated with daily PM. Studies found an average 1.4% increase in cardiovascular disease with a 10 $\mu g/m^3$ increase in $PM_{10}$. Although this percentage increase is lower than the percentage increase associated with respiratory disease (3-4%), it represents many more individuals since cardiovascular deaths are much more common than deaths from respiratory disease. A recent study reported a positive association between $PM_{10}$ concentrations and odds of pulse rate being elevated by 5 to 10 beats per minute in elderly subjects in Utah ((Pope et al, 1999). Contrary to expectation, there was no significant association between $PM_{10}$ and arterial oxygen saturation.

## CHRONIC EFFECTS

Most of the studies reviewed so far have evaluated the short-term effects of PM exposure and health outcomes. Designing a study to evaluate the long-term or chronic effects of PM exposure is more difficult because one loses the temporality of daily associations and because of all the other factors that can affect health longitudinally. One cluster of studies, the California Seventh-Day Adventist study, has followed a group of subjects since 1973 (Abbey et al, 1995). This particular group of subjects is valuable since, as a group, they do not smoke or use alcohol, thus removing two potential confounding factors. The latest analysis of health outcomes in this group found that $PM_{10}$ had a significant association with development of airway obstructive disease, with chronic productive cough, and with increased severity of asthma. The relative risk of development of these disorders was greater for subjects who had a history of exposure to higher concentrations of $PM_{10}$. Exposure was based on both the concentration of $PM_{10}$ and the duration of estimated exposure as seen in Figure 10-7.

Another study examined air pollution from truck traffic and lung function in children living near motorways in the Netherlands (Brunekreef et al, 1997). Air pollution levels were estimated by separate traffic counts for automobiles and trucks. Lung function in children was associated with traffic; the association was strongest for children living within 300 meters of the roadway. Associations were stronger for females than males and more strongly associated with truck traffic than automobile traffic, which may suggest an adverse effect from diesel exhaust.

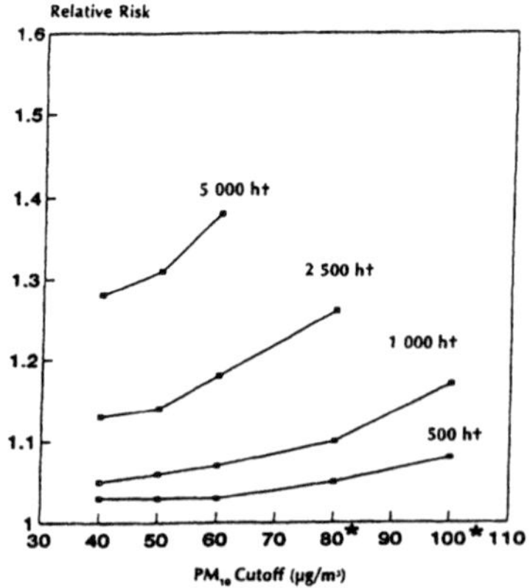


*Figure 10-7. Relative risk of development of new cases of definite symptoms of obstructive airway disease for differing average hours in excess of 40, 50, 60, 80, and 100 $\mu g/m^3$ $PM_{10}$. * Statistically significant. + Average annual hours from 1973-1977. FROM: Abbey et al, 1995.*

Another study evaluated the long-term effects of exposure to air pollution by studying histopathologic damage in the lungs of victims of sudden traumatic deaths in Brazil (Souza et al, 1998). One indicator of air pollution exposure was number of spots of carbon deposition in lymphatic drainage areas. Information on smoking history was obtained from family members. Lungs collected in areas of high air pollution presented more evidence of histopathologic damage than did lungs collected in an area of lower air pollution.

## POLICY CONCERNS AND RESEARCH NEEDS

The new $PM_{2.5}$ standard is extremely controversial, with industry claiming the science does not back up the standard and that the economic cost is too high, while the health community claims that the 24 hour standard (65 $\mu g/m^3$) will not represent a tightening of PM pollution in many cities. Throughout the history of the CAA, questions have been raised about whether the health benefits implicit in compliance with NAAQS could justify the costs incurred by industry and taxpayers. In 1990 Congress directed EPA to conduct a study to assess this concern. Shortly after the new standard was announced, EPA released the summary of a year-long study on the benefits and costs of the Clean Air Act since its inception in 1970 (EPA, 1997). The committee determined that the benefits of the CAA from 1970-1990 ranged from 5.6 to 49.4 trillion dollars with an estimate of an average of $22.2 trillion. Over the same time period the estimated cost of compliance with the CAA standards was approximately 0.5 trillion dollars. Thus, the estimated benefit to US citizens is approximately 21.7 trillion dollars. Obviously estimating these costs was difficult, but since the benefit was over 20 times the cost there really is not any argument as to whether it is economically feasible to clean up our air. Based on the concerns about the scientific basis for the new $PM_{2.5}$ standard, the US Congress asked the National Research Council (NRC) to review the research needs in the area of particulate matter health effects. A list of the top ten topics determined by the NRC is given in table 10-6. The list represents a 14-year research agenda for EPA (NRC. 1998).

*Table 10-6. Top ten topics for particulate matter research*

*Investigate quantitative relationships between PM and individual exposures*
*Assess the most biologically important characteristics of PM*
*Identify the most susceptible populations*
*Analyze exposure to the most biologically important constituents*
*Develop advanced mathematical, modeling, and monitoring tools.*
*Apply modeling to link sources to exposed individuals*
*Investigate deposition patterns and fate of particles*
*Analyze interactions between PM and gaseous co-pollutants*
*Explore underlying mechanisms that explain the health effects*
*Develop advanced methods for statistical analysis of epidemiologic studies*

*FROM: National Research Council, 1998.*

The lack of a list of toxicologic mechanisms to explain the health effects seen in epidemiologic studies is a primary criticism directed at setting a NAAQS for $PM_{2.5}$. Unlike the scientific literature for ozone, where there are hundreds of animal studies clearly showing mechanisms of oxidative effects, there is a real paucity of animal studies exploring mechanisms of PM. The primary reason for this is the extreme difficulty of creating a test atmosphere in a laboratory that resembles outdoor PM. Since PM is a complex mixture that varies from region to region, from season to season, and from daytime to nighttime, generating PM in a laboratory is a real challenge. Several years ago a new approach to generating PM was developed using a virtual impactor (Sioutas et al, 1995). This new technique involves a method to concentrate outdoor PM to levels of 10-fold higher than ambient levels. These concentrated air particles (CAPs) remain airborne and thus can be used for animal and human exposures. Animals exposed to PM in this fashion have shown some of the adverse health effects suggested by epidemiologic studies. Dogs exposed to the CAPs developed cardiac dysrhythmias. Also, when comprised animals (those manipulated to exhibit disease which mimics chronic human respiratory or cardiac disease) are used in PM exposure studies, they show increased mortality after simulated PM exposures compared to healthy animals. The toxicologic effects of PM are summarized in the latest criteria document on PM compiled by EPA (1996). The NRC report and other publications have developed a list of hypotheses for PM related health effects. This list, not ordered by priority, is given below.

*Table 10-7. Current hypotheses of PM characteristics related to health effects.*

*PM mass concentration*
*Metals, especially soluble metals*
*Acidity*
*Organic compounds*
*Ultrafine PM*
*Biologic material*
*Sulfates and nitrates (Secondary aerosols)*
*Peroxides*
*Elemental carbon (soot)*
*Co-pollutants*

---

There are data that relate to some of these hypotheses.

PM mass concentration: PM mass concentration is the metric that has been used in most studies to date, since that has been the approved method for monitoring. A recent summary of the goodness of fit with mortality data of three size fractions of PM was published by Schwartz and associates (1997) and discussed earlier. A summary of those data is given in Table 10-4. Many studies now have been conducted evaluating the effects of PM with differing concentrations of various metals such as soluble iron, vanadium, and nickel. It is hypothesized that soluble iron when inhaled may activate the Fenton reaction leading to the production of more free radicals in the lung. One study found a significant increase in expression of inflammatory cytokines following exposure of human respiratory cells to a form of PM air pollution, residual oil fly ash (ROFA). When the cells were treated with a chelating agent (which would bind metal and prevent them from interacting with the cells), the significant increase in the cytokines disappeared (Carter et al, 1997). Data from that study are given in Figure 10-8.

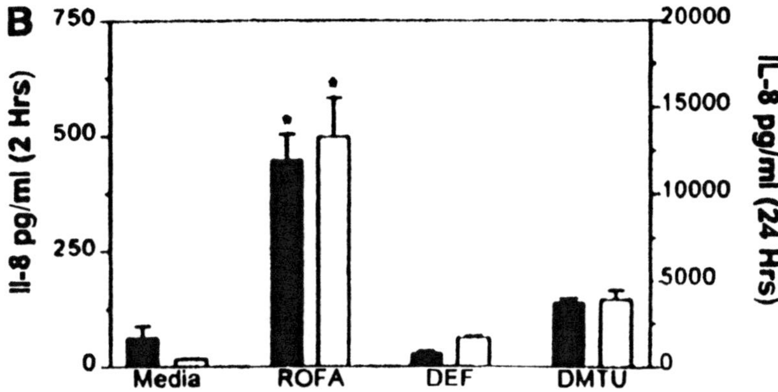


*Figure 10-8. Inhibition of ROFA-induced IL-8 protein secretion by deferoxamine, a chelating agent. Dark bar = 2 hr exposure; light bar = 24 hr exposure. Similar results were seen for IL-6. FROM: Carter et al, 1997 with permission.*

Acidity: The effects of acidity (in the form of sulfuric acid) have been fully documented in studies with rabbits and donkeys (as well as some human studies), These data are discussed in Chapter 9 which describes the health effects of sulfur oxides.

Organic compounds: It is known that organic compounds in ambient air attach to PM particles and thus are deposited into the lung. Many of these compounds have known toxicities such as polyaromatic hydrocarbons (PAHs).

Ultrafine PM: Filter sampling of PM in ambient air measures the mass of the particles in the air. There is a hypothesis that rather than mass being the toxic marker it is the number of particles in the air that are responsible for toxicity. Measuring the number of particles is much more difficult than simply measuring mass and few data exist. Refer back to the figures on size mode of particles in Ch 4, Figure 4-4a and 4-4b. Particles below roughly 0.30 micrometers are referred to as ultrafine (UF) particles. There are instruments available which can measure these UF particles which range from 40 to 300 nanometers. Investigators have tried to determine the relative toxicity of UF particles. One study found increased mortality in rats exposed to UF particles compared to particles in the range of 2.5 micrometers (Oberdorster et al, 1997).

Biologic material: Studies of asthma and pollen address some of the concerns regarding the importance of biologic material. Tree and grass pollens are known to aggravate asthma. A study by Anderson and associates (1998) in London had daily data on grass, oak, and birch pollen. They found that ozone, $SO_2$, $NO_2$, and PM were all weakly associated with daily hospital admissions for asthma and that there was no evidence for confounding by pollen. However a study in Mexico City came to the opposite conclusion; they did find that grass pollen and fungal spores were statistically associated with asthma hospital admissions more strongly than were air pollutants (Rosas et al, 1998). The relationship was stronger during the dry season than doing the wet season. Pollen counting and identification is extremely time consuming and requires expertise. Nevertheless it is very important to resolve the issue of the independent effects of pollens and air pollutants.

## CONCLUSION

This review of the health effects of PM should have given readers an idea of the excitement present in the field among investigators who are conducting pertinent research. EPA already is in the process of preparing another criteria document for PM (as dictated by the 5-year cycle for reviewing NAAQS). Some investigators are scrambling to develop more precise statistical models for testing association between health outcomes and PM while others are conducting studies of personal exposure to PM in sensitive populations in order to provide more accurate exposure assessments. Evaluating the cardiac health effects of PM is one very active area of research. EPA is funding sites around the US that will analyze the chemical composition of PM thus providing a basis for toxicologic studies. No doubt a review of the health effects of PM in 5 to 10 years will look much different than the summary presented here.

# REFERENCES

Abbey DE, Hwang BL, Burchette RJ, et al. Estimated long-term ambient concentrations of $PM_{10}$ and development of respiratory symptoms in a nonsmoking population Arch Environ Health 1995; 50: 139-152.

Anderson HR, Ponce de Leon A, Bland JM, et al. Air pollution, pollens, and daily admissions for asthma in London 1987-1992. Thorax 1998; 53: 842-848.

Anto JM, Sunyer J Rodriguez-Roison R, et al. Community outbreaks of asthma associated with inhalation of soybean dust. N Engl J Med 1989; 320: 295-300.

Bates DV, Baker-Anderson M, Sizto R. Asthma attack periodicity: A study of hospital emergency visits in Vancouver. Environ Res 1990; 51: 51-70.

Bates DV. Health indices of the adverse effects of air pollution: The question of coherence. Environ Res 1992; 59: 336-349.

Brunekreef B, Janssen NAH, de Hartog J, et al. Air pollution from truck traffic and lung function in children living near motorways. Epidemiology 1997; 8: 298-303.

Burnett RT, Cakmak S, Brook JR, Krewski D. The role of particulate size and chemistry in the association between summertime ambient air pollution and hospitalization for cardiorespiratory diseases. Environ Health Perspect 1997; 105: 614-620.

Carter JD, Ghio AJ, Samet JM, Devlin RB. Cytokine production by human airway epithelial cells after exposure to an air pollution particle is metal-dependant. Toxicol Appl Pharmacol 1997; 146: 180-188.

Diaz-Sanchez D, Tsien A, Casillas A, et al. Enhanced nasal cytokine production in human beings after in vivo challenge with diesel exhaust particles. J Allergy Clin Immunol 1997; 98: 114-123.

Dockery DW, Speizer FE. Epidemiological evidence for aggravation and promotion of COPD by acid air pollution. Lung Biol Health Dis 1989; 443: 201-225.

Dockery DW, Speizer FE, Stram DO, et al. Effects of inhalable particles on respiratory health of children. Am Rev Respir Dis 1989; 139:587-594.

Dockery DW, Pope CA III. An association between air pollution and mortality in six US cities. N Engl J Med 1993; 329: 1753-1759.

Dockery DW, Pope CA III. Acute respiratory effects of particulate air pollution. Ann Rev Public Health 1994; 15: 107-132.

EPA. The benefits and costs of the Clean Air Act. October 1997.

Fairley D. The relationship of daily mortality to suspended particulates in Santa Clara county, 1980-1986. Environ Health Perspect 1990; 89: 159-168.

Gerrity TR, Lee PS, Hass EJ, et al. Calculated deposition of inhaled particles in the airway generations of normal subjects. J Appl Physiol 1979; 47: 867-873.

Janssen NAH, Hoek G, Brunekreef B, et al. Personal sampling of particles in adults: relation among personal, indoor, and outdoor air concentrations. Am J Epidemiol 1998; 147: 537-547.

Koenig JQ, Larson TV, Hanley QS, et al. Pulmonary function changes in children associated with fine particulate matter. Environ Res 1993; 63: 26-38.

Knox RB, Suphioglu C, Taylor P, et al. Major grass pollen allergen Lol p 1 binds to diesel exhaust particles; implications for asthma and air pollution. Clin Exp Allergy 1997; 27: 246-251.

Larson TV, Koenig JQ. Wood smoke: Emissions and noncancer respiratory effects. Ann Rev Public Health 1994; 15: 133-156.

Leclair V. California considers tougher diesel exhaust rule. Environ Sci Technol 1997; 31: 355A.

Moolgavkar SH, Luebeck EG. A critical review of the evidence on particulate air pollution and mortality. Epidemiology 1996; 7: 420-428.

National Research Council. Research priorities for airborne particulate matter; I. Immediate priorities and a long-range research portfolio. National Research Council, Washington DC, March 1998.

Nel AE, Diaz-Sanchez D, Ng D, et al. Enhancement of allergic inflammation by the interaction between diesel exhaust particles and the immune system. J Allergy Clin Immunol 1998; 102: 539-554.

Oberdorster G, Osier M. Intratracheal inhalation vs intratracheal instillation: differences in particle effect. Fund Appl Toxicol 1997; 40: 220-227.

Ostro BD, Lipsett MJ, Wiener MB, Selner JC. Asthmatic responses to airborne acid aerosols. Am J Public Health. 1991; 81: 694-702.

Periz-Padilla R, Regalado J, Vedal S, et al. Exposure to biomass smoke and chronic airway disease in Mexican women. Am J Respir Crit Care Med 1996; 154; 701-706.

Peters A, Goldstein IF, Beyer U, et al. Acute health effects of high levels of air pollution in eastern Europe. Am J Epidemiol 1997; 144: 570-581.

Ponka A. Asthma and low level air pollution in Helsinki. Arch Environ Health 1991; 46: 262-270.

Pope CA III, Dockery DW, Spengler JD, Raizenne ME. Respiratory health and $PM_{10}$ pollution: A daily time series analysis. Am Rev Respir Dis 1991; 144: 668-674.

Pope CA III. Respiratory hospital admissions associated with $PM_{10}$ pollution in Utah, Salt Lake, and Cache Valleys. Arch Environ Health 1991; 46: 90-97.

Pope CA III, Schwartz J, Ransom, MR. Daily mortality and $PM_{10}$ pollution in Utah valley. Arch Environ Health 1992; 47: 211-217.

Pope CA III, Thun MJ, Namboodiri MM, et al.. Particulate air pollution as a predictor of mortality in a prospective study of U. S. adults. Am Rev Respir Crit Care Med 1995; 151: 669-674.

Pope CA III, Dockery DW, Kanner RE, et al. Oxygen saturation, pulse rate, and particulate air pollution. Am J Respir Crit Care Med 1999; 159: 365-372.

RosasI, McCartney HA, Payne RW, et al. Analysis of the relationships between environmental factors (aeroallergens, air pollution and weather) and asthma emergency admissions to a hospital in Mexico City. Allergy 1998; 53: 394-401.

Schwartz J, Spix C, Wichmann HE, Malin E. Air pollution and acute respiratory illness in five German communities. Environ Res 1991; 56: 1-14.

Schwartz J, Dockery DW. Increased mortality in Philadelphia associated with daily air pollution concentrations. Am Rev Respir Dis 1992; 145: 600-604.

Schwartz J, Dockery DW. Particulate air pollution and daily mortality in Steubenville, Ohio Am J Epidemiology 1992; 135: 12-19.

Schwartz J, Koenig JQ, Slater D, et al. Particulate air pollution and hospital emergency room visits for asthma in Seattle. Am Rev Respir Dis 1993; 147: 826-831.

Schwartz J, Dockery DW, Neas LM. Is daily mortality associated specifically with fine particles? J Air Waste Manage Assoc 1996; 46: 927-939.

Sheppard L, Levy D, Norris G, Effects of ambient air pollution on nonelderly asthma hospital admissions in Seattle, Washington, 1987-1994. Epidemiology 1999; 10: 23-30.

Silverman F, Hosein HR, Corey P, et al. Effects of particulate matter exposure and medication use on asthmatics. Arch Environ Health 1992; 47: 51-56.

Sioutas C, Koutrakis P, Ferguson ST, Burton RM. Development and evaluation of a prototype ambient particle concentrator for inhalation exposure studies. Inhal Toxicol 1995; 7: 633-644.

Souza MB, Saldiva PHN, Pop CA III, Luiza Capelozzi V. Respiratory changes due to long-term exposure to urban levels of air pollution. Chest 1998; 113: 1312-1318.

Vander AJ, Sherman JH, Luciano DS. Human Physiology: The mechanisms of body function., McGraw-Hill, New York, 1970. p 303.

Zhu X. Development of a potential biomarker of environmental wood smoke. Masters Thesis. Department of Environmental Health, University of Washington, Seattle, 1998.

# CHAPTER 11.

# HEALTH EFFECTS OF OZONE

Ozone is the most persistent, intractable air pollutant in urban air. In 1995, according to EPA, over 70 million people in the US resided in areas not meeting the one-hour EPA ozone standard. As mentioned earlier, the adequacy of the US EPA NAAQS standard for ozone was severely criticized causing the agency to review the standard and promulgate a new standard in 1997. The previous standard was 0.12 ppm averaged over 1 hour. The new standard is 0.08 ppm averaged over 8 hours. This new standard reflects current ozone patterns more closely. Years ago, ozone concentrations peaked 2-3 hours after morning and afternoon commuting traffic and those peaks were of about 1 hour in duration. However traffic patterns have changed substantially in most US cities, and now ozone concentrations are seen to last for approximately 8 hours after first rising around 10 am. The reader should keep the old and new standards and their averaging times in mind as she or he reads the following summary of the health effects of ozone exposures.

Recall that ambient ozone is what is called a secondary pollutant, meaning that rather than being directly emitted as ozone, it is formed in the atmospheric from nonmethane hydrocarbons in the presence of nitrogen oxides and ultraviolet light (see Ch 4 for details). Table 4-8 shows a schematic for formation of ozone in urban ambient air. Thus communities that attempt to reduce ozone levels must direct their regulatory activity at the precursors, both hydrocarbons and $NO_x$. Another consequence of the complicated atmospheric chemistry resulting in ambient ozone in urban areas is that the higher ozone concentrations are not where the precursors are highest but are 20-40 kilometers downwind from the precursors reflecting the interval between precursor emissions and photochemical transformation into ambient ozone. This disconnect is also driven by the fact that nitric oxide (NO), at certain concentrations, can eat up or scavenge ozone, so that ozone levels are never their highest close to sources of $NO_2$. This fact has consequences for the study of health effects of ozone. For the year 1997, the 8-hour maximum concentration of ozone was 0.09 ppm at the downwind monitor in the photochemical belt (transport corridor) and 0.04 ppm at a monitor in the city of Seattle. This presents a challenge for an epidemiologist who wants to test for associations between health endpoints and ozone for residents of the greater Seattle area. What is the appropriate ozone concentration for use in such a study?

Ozone is an ideal molecule to cause destructive effects to the respiratory system. Ozone is an oxygen molecule with one additional oxygen atom changing oxygen ($O_2$) to ozone ($O_3$). However the extra atom is not bound tightly to oxygen and easily breaks off. Upon inhalation, the extra ozone atom will break off at the first encounter with any biologic tissue. Usually this site is the epithelial lining of the nasal passages or the bronchial airways. The epithelial lining fluid (ELF) provides the first line of defense against inhaled toxicants. The ELF is especially effective in defending against oxidant gases such as ozone as it contains many antioxidant substances. Because the thickness of the ELF varies throughout the airways, $O_3$ reactivity most likely includes cell membrane penetration as well as ELF components (Pryor, 1992). Research shows that ozone travels less than 0.1 micron across tissue in the lung (Pryor, 1992). However the singlet oxygen atom created in the breakdown of ozone ($O_3$) to oxygen ($O_2$) can penetrate to the cellular level. This singlet oxygen atom is referred to as a free radical and it is available to enter into various oxidizing reactions. As this happens there can be adverse reactions for the membranes, enzymes, and other biological systems which suffer the free radical effect. Antioxidant defenses in the lung defend our tissues from free radical (or other oxidant) damage. Humans also breathe (and depend upon) oxygen which also is an oxidant. Also some ozone is formed from vegetative sources such as terpines emitted by trees, so low concentrations of ozone always have been a natural part of the human environment. Thus as we inhale ozone during our everyday activities, there is a balance between protection provided by our bodies antioxidant defenses and possible ozone-induced damage should those defenses be overwhelmed. A list of important antioxidants known to be present in the ELF is given in Table 11-1. The lining fluid is very thin, ranging about 1-2 μm in the bronchi and less than 0.5 μm in the alveolar sacs. Thus it is not surprising that the site of ozone injury often seen in animal studies is the alveolar region. Oxidant injury to the tissue by oxidants such as ozone inhalation triggers an antioxidant defense. A schematic of a possible pathway for ozone from the airstream to underlying cells is given in Figure 11-1. A recent study investigated the effect of ozone exposure on ELF derived from human BAL in an in vitro system. Ozone exposure, as low as 50 ppb, significant reductions of ascorbic acid, uric acid, and glutathione (Mudway et al, 1996).

*Table 11-1. Common antioxidants present in epithelial lining fluid*

| |
|---|
| *Glutathione* |
| *Ascorbic acid (Vitamin C)* |
| *uric acid* |
| *Bilirubin* |
| *α-tocopherol (Vitamin E)* |
| *β-carotene* |
| *ubiquinol-10* |
| *Albumin-SH* |

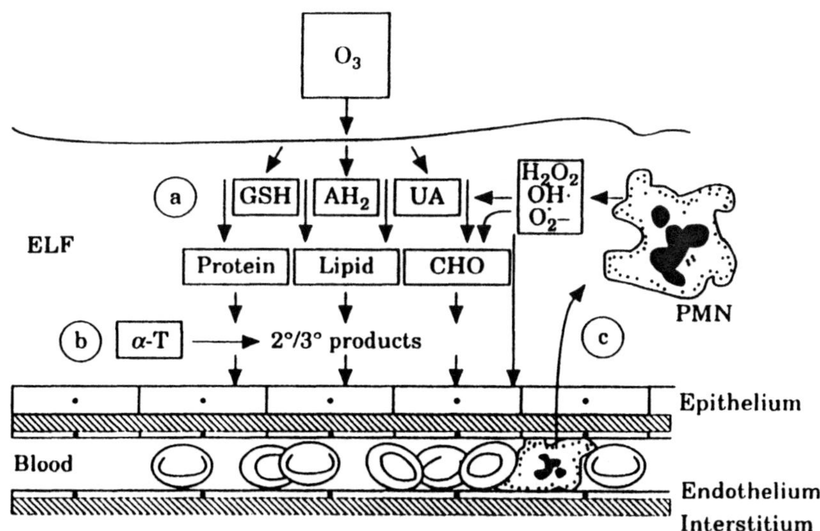


*Figure 11-1. Potential mechanisms of ozone-induced injury in the lung. GSH = glutathione, $AH_2$ = ascorbate; UA = urate; α-T = α-tocopherol; CHO = carbohydrate; $H_2O_2$ = hydrogen peroxide; PMN = neutrophil; FROM: Kelly et al, 1995 with permission.*

One major mechanism of $O_3$ toxicity is through oxidant damage. Ozone reacts with polyunsaturated fatty acids in vitro to form free radicals (Menzel, 1992; Pryor and Church, 1991). There is considerable question however, regarding how much $O_3$ reaches cell membranes in the lung, since $O_3$ first encounters the lung epithelial lining fluid (ELF) and reacts with water soluble antioxidants such as GSH and ascorbate, as well as glycoproteins and unsaturated fatty acids. Regardless of the precise target of $O_3$, ozonation products such as lipid free radicals exert a powerful oxidant burden on the lung that results in depletion of both extracellular and intracellular antioxidant defenses as well as a cascade of secondary and tertiary products including lipid peroxides and reactive oxygen species from neutrophil infiltration (Kelly, 1995).

Since some antioxidant agents are influenced by diet, there is increasing interest in relationships between plasma (and assumed lung) levels of antioxidants and lung function. In general, it has been shown that dietary antioxidant vitamin intake is associated with normal values of lung function and related to protecting against the development of chronic obstructive lung disease (Britton et al, 1995). Individuals with asthma have lower antioxidant plasma levels than the population at large (Hatch, 1995). In reviews examining the role of oxidant stress in the development of asthma, low dietary intakes of vitamins C and E were linked with probable increases in the risk of developing asthma. A prospective study of diet and adult-onset asthma based on the Nurses' Health Survey found an inverse relation between dietary vitamin E and 6-year-risk of asthma (Troisi et al, 1995).

It is one of the tenets of risk assessment that, as well as determining whether a given agent is toxic, it is necessary to determine whether there also is exposure to human populations and to vulnerable target tissues. As discussed earlier there is wide spread exposure to ozone in the US. Measurements of uptake of gases by the respiratory system are difficult to achieve. Ozone uptake became quantifiable when direct measuring instruments with a rapid response time were developed. In order to measure uptake of a gas it is necessary to be able to measure the concentration in inspired air and then subsequently in exhaled air. Thus the ozone analyzer has to have a reaction time in the range of 1-2 seconds. In the 1970s it was assumed that very little ozone deposited in the upper airway due to the fact that ozone is not very water soluble. However once the measurements were possible, a different story appeared. Uptake studies have determined that approximately 40% of inhaled ozone is taken up in the upper airway (nasal passages). Total uptake by the time the inhaled air reaches the lower airways is 91%. These values are dependent on ventilation rate and the above values were recorded during quiet breathing. Significantly less ozone is taken up at a ventilatory rate of 24 breaths/min than at 12 br/min (Gerrity et al, 1988; 1994).

There are three fairly recent reviews of the health effects of ozone in human populations (Lippmann, 1992; US EPA, 1996; Bylin et al, 1996). Another recent review is restricted to animal data but offers an excellent summary of morphological and biochemical ozone-induced effects (Chitano et al, 1995). The Chitano review summarizes lung function and airway responsiveness studies in several animal species. Functional alterations have been reported after ozone exposure at concentrations as low as 0.24 ppm. However we will see later in this chapter that pulmonary function decrements in humans have been recorded after 0.08 ppm ozone exposure. The animal research emphasizes the importance of ozone effects mediated by the nervous system, primarily from the vagus nerve whose activity tends to increase respiratory rate. Rapid, shallow breathing is one hallmark of ozone exposure. Actually, human exposures to ozone result in neurally mediated pain upon deep inspiration as will be described in more detail later in this chapter. Bronchial hyperresponsiveness (BHR) following ozone exposure has been shown with histamine and acetylcholine challenges in animals. BHR is also a consequence of ozone-induced effects in human subjects. The animal studies allow investigators to experiment with various agents that may inhibit ozone-induced BHR. Methods for inhibiting BHR include administration of inhibitors of arachodonic acid metabolism, namely indomethacin and experimental cyclooxygenase inhibitors. Atropine also has been studied as a possible mitigator of ozone induced effects. The effects of pharmaceutical agents and antioxidants on ozone-induced respiratory effects in humans will be discussed later. Animal studies also allow examination of histopathologic and structural changes that cannot be assessed in human subjects. In general, ozone exposure in animals induces cell injury, lung edema, and inflammation. Structural damage is often seen at the junctions of the terminal airways. Interestingly this is the place in the airways where the mucus lining becomes spotty and soon disappears altogether and where the ELF is thinnest. Animal studies clearly show that duration of exposure is a determining factor in the degree of ozone-induced lung damage. The discussion of structural changes in this review is useful in helping investigators of human effects understand the likely areas of maximal effect within the tracheobronchial tree and alveolar spaces. Long term exposure to ozone has been associated with fibrotic changes resulting is a smaller, stiffer lung. Animal researchers also have studied repair of ozone-induced damage. In fact, in some studies signs of repair begin to occur during exposure. Tolerance to repeated and continuous ozone exposure has been shown in animal studies, and, as discussed later, also in human exposures.

As pointed out in Ch 7, extrapolation from results in experimental animals who have very different lung structures to human populations is difficult. Another complicating factor, especially for interpretation of ozone effects, is the remarkable difference in ELF composition among species. Levels of certain antioxidants in guinea pigs and rats compared to humans are given in Table 11-2.

*Table 11-2. Comparison of estimated levels (μM) of certain antioxidants in ELF from different mammals. Adapted from Cross et al, 1994.*

| | *Guinea pig* | *Rat* | *Human* |
|---|---|---|---|
| *Ascorbic acid* | *600* | *2000* | *160* |
| *Glutathione* | *220* | *130* | *165* |
| *Uric acid* | *53* | *46* | *129* |
| *α-Tocopherol* | *0.8* | *75* | *2.5* |

In vitro studies using human respiratory cells retrieved from airway lavage make a nice blending of the type of control one achieves in an animal study and the relevance of the use of human biology. In vitro studies of nasal and bronchial epithelial cells and also alveolar macrophages either primary cells or in cell lines are an interesting addition to the health effects of air pollution literature.

## UPPER AIRWAY EFFECTS

All of the earliest research on ozone effects in human subjects was conducted using the lower respiratory tract as the endpoint of concern. Most of the early research was restricted to reporting pulmonary function changes. Once it had been shown that ozone actually was taken up in the nasal passages it became important to determine the physiological effect of such uptake. The first study to explore this question measured ozone effects in the nose of healthy subjects using the nasal lavage technique that is described in Chapter 7. In that study, Graham and co- workers (1988) found increased neutrophils in nasal lavage fluid after 2 hour exposure at rest to 0.4 ppm $O_3$. A subsequent study investigated ozone effects in subjects with and without asthma, again using nasal lavage technique (McBride et al, 1994). Increased neutrophils were seen in asthmatic but not nonasthmatic subjects after 0.24 ppm $O_3$ exposure for 90 minutes during intermittent exercise. Calderon-Garciduenas et al. (1992; 1993) documented nasal pathology associated with ozone exposure in residents of Mexico City. When individuals move to Mexico City from less polluted, rural areas, pathological changes appear in nasal mucosa within two to three weeks. Thus there are well-documented effects of ozone on nasal endpoints.

## LOWER AIRWAY EFFECTS

Ozone has been studied in controlled laboratory settings more carefully than the other criteria pollutants. The following summary of lower respiratory effects is divided into three categories: pulmonary function changes; measures of bronchial hyperresponsiveness (BHR); and measures of airway inflammation. Studies have been conducted with healthy young adults, healthy children, healthy senior subjects, subjects with asthma, smokers, and different ethnicity. Nevertheless, identity of the most susceptible group or groups with respect to ozone exposure is not clear.

### Pulmonary Function Changes

**Healthy Young Adults**

Exposure to ozone in controlled laboratory settings has been associated with decrements in lung function in a large number of studies. The lung function changes are both time and dose dependent. Duration appears to be the more important of these two parameters. Significant decrements in $FEV_1$ have been shown in healthy subjects inhaling 0.08 ppm ozone for 6.6 hours. These changes at various ozone concentrations are shown in Figure 11-2. The reader should note that these changes are seen at a concentration and averaging time essentially the same as the new revised 1997 NAAQS for ozone. Also these changes were found in healthy subjects, whereas much of the concern for the adequacy of the air pollution standards has been protection of susceptible populations.

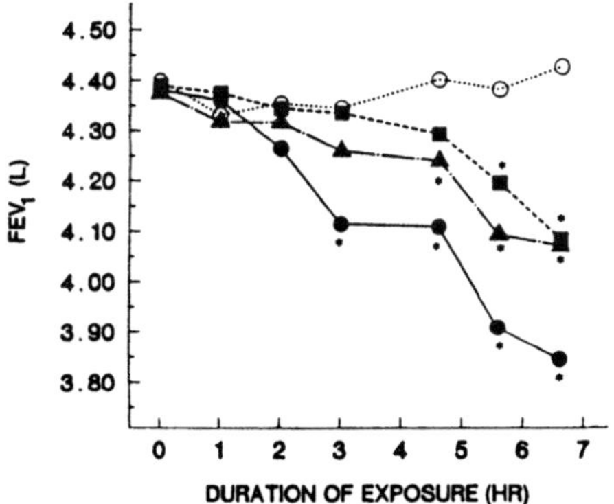


*Figure 11-2. Decreases in $FEV_1$ per hour exposure at air and three concentrations of ozone. *Open circles = air; squares = 0.08 ppm; triangles = 0.12 ppm; solid circles = 0.16 ppm. FROM: Horstman and Folinsbee, 1990 with permission.*

Decrements in pulmonary function associated with ozone exposure have been seen in real life settings, as well as in the controlled laboratory setting. A study evaluated pulmonary function in adult hikers during one summer season and found a 2.6% decline in $FEV_1$ for each 0.05 ppm increase in ambient ozone concentration (Korrick et al, 1998).

**Inter-individual Variability**

One of the interesting, recurring results seen in studies of ozone exposure in healthy young adults is the wide range in variability in the degree of pulmonary function effect. In general, changes in lung function following controlled ozone exposures reveal that approximately 10-20% of healthy subjects participating in these studies experience a pronounced response to ozone. Frequencies of $FEV_1$ decrement in healthy adult subjects after exposure to various ozone concentrations are given in Table 11-3.

*Table 11-3. Frequency of $FEV_1$ decrement in young adults after ozone exposure for 2 hours during intermittent exercise.*

| *Ozone concentration* | *$FEV_1$ decrements* | *N (subjects)* |
|---|---|---|
| *0.00 ppm* | *10-20%* | *0/20* |
| | *> 20%* | *0/20* |
| *0.12* | *10-20%* | *3/22* |
| | *> 20%* | *0/22* |
| *0.18* | *10-20%* | *2/20* |
| | *> 20%* | *2/20* |
| *0.24* | *10-20%* | *3/21* |
| | *> 20%* | *7/21* |
| *0.30* | *10-20%* | *6/20* |
| | *> 20%* | *6/20* |
| *0.40* | *10-20%* | *7/20* |
| | *> 20%* | *13/20* |

*Adapted from McDonnell et al, 1983*

There is great interest in understanding the basis for this differential response. Some of it must be due to underlying bronchial hyperresponsiveness (BHR) and whether this is genetically determined is not known. Also as discussed later, diet probably plays a role in BHR. Subjects usually are screened for BHR and in general the wide variation has been seen in subjects who do not appear to have BHR. However there probably are a few subjects who are missed with screening techniques. Nevertheless there must be some other risk factor not yet identified. Ozone effects on BHR are discussed later.

**Reproducibility**

Several investigators have exposed the same individuals on repeated occasions to the same concentrations of ozone and found that, in general, there is good intra-subject reproducibility. Subjects sensitive to ozone on the first exposure are sensitive the next time and similarly those not responsive remain not responsive as long as the conditions are constant.

**Adaptation**

Adaptation also has been reported in controlled studies; when subjects are exposed to a given concentration of ozone on 5 consecutive days, there is an average decrement in lung function on day 1, a larger decrement on day 2 and 3, a smaller decrement on day 4, and on day 5 the average lung function values are no different from baseline values recorded prior to the first day of ozone exposure. The reason for the attenuation of pulmonary function effects is not known. There are suggestive data that indicate that ozone-induced inflammatory changes do not attenuate with repeated exposure. Also there apparently is some type of adaptation that occurs when individuals are exposed frequently during the ozone season. One study found that a group of subjects had no consistent pulmonary response to ozone in a controlled laboratory setting during the summer but that, when these same subjects were exposed to the identical conditions in late winter, a significant pulmonary function decrement was elicited (Hackney et al, 1989). This finding was interpreted as showing the subjects had adapted to ozone in ambient air during the summer and thus were non-responsive to ozone in the controlled setting.

**Children**

Very little research has been carried out using controlled laboratory exposures in children under the age of 12-13. Epidemiologic studies of children at summer camps in areas with significant air pollution including ozone are discussed later in the section on epidemiology. One group of investigators evaluated the response to ozone in 33 boys and girls aged 8-11 in a simulated outdoor setting in Los Angeles (Avol et al, 1987). In that study, subjects were exposed for one hour to either purified air or ambient air drawn into the mobile laboratory. The average ambient concentration of ozone was 0.113 ppm (slightly below the old national ambient air quality standard of 0.120 ppm for one hour). As a group, the subjects showed no significant reductions in pulmonary function. Another study with this same age group also found no pulmonary function effects, however those authors did note that the children failed to record respiratory symptoms during ozone exposures (McDonnell et al, 1985). Lack of symptom reporting causes some concern, since symptoms of shortness of breath or pain on inspiration are considered hallmarks of ozone exposures and may alert subjects to discontinue exercise.

**Subjects with Asthma**

There are studies that indicate that subjects with asthma are more sensitive to the inhaled effects of ozone than subjects without asthma. A summary of these studies has been published (Koenig, 1995). Silverman (1979), almost two decades ago, demonstrated an exaggerated pulmonary function response to a 2-hour exposure to ozone at rest. Hackney and co-workers (1989) found that when 59 subjects were screened for ozone sensitivity, the ozone reactive subjects all had a history of allergy, mild asthma, or previous evidence of sensitivity to ozone in a previous exposure setting. In the study cited earlier showing decrements in pulmonary function in adults hikers, it was seen that subjects in that study who had asthma showed a four-fold greater decrement in responsiveness to ozone (Korrick et al, 1998). Aris and co-workers ((1991) found that subjects who qualified for a study based on sensitivity to ozone all had a positive response to a methacholine challenge. The methacholine challenge, as described in Chapter 7, is a test for BHR, one of the signs of asthma.

#### Age as a Risk Factor for Sensitivity to Ozone

In the 1980s there was interest in documenting whether older individuals, even though they were healthy, were more susceptible to inhaled ozone. The results actually turned out the other way; most studies have found that subjects over the age of 60 seem to show less of a decrement in lung function following ozone exposure than do young adults (see, for instance, Drechsler-Parks et al, 1987). In fact, there is an indication that younger individuals are more responsive to ozone. Data were analyzed from 290 white male subjects aged 18-30 exposed to ozone ranging from 0.12 - 0.40 ppm or clean air. There was a significant association between age and $FEV_1$ decrements after ozone exposure; younger subjects showed greater decrements in $FEV_1$. Age predicted 4% of the decrement in $FEV_1$ (see Figure 11-3). Although these data examining the relationship between age and ozone response have been conducted on subjects without asthma, it is interesting to speculate that, if younger healthy subjects are at a slight increased risk with respect to ozone exposure, adolescent subjects with asthma may be even more at risk (Koenig, 1998).

#### Subjects with Chronic Obstructive Pulmonary Disease (COPD)

Although older healthy subjects appear to be no more or even less sensitive to the inhaled effects of ozone than their younger counterparts, older subjects with chronic disease have been studied rarely. One study of nine male subjects with severe COPD matched with 10 healthy males in the same age range found an average 19% decrement in $FEV_1$ in the subjects with COPD compared with an average 2% decrement in the healthy control subjects (Gong et al, 1997). These subjects were exposed to 0.24 ppm ozone or air for 4 hours during intermittent light exercise.

**Smokers**

Exposure to ozone in subjects who are chronic cigarette smokers has resulted in a paradoxical finding. It was assumed that smokers would be more vulnerable to the effects of inhaled ozone due to their state of constant airway irritation. However, experiments have shown, that opposite to what was expected, cigarette smokers in general do not show pulmonary function decrements at concentrations known to produce that effect in healthy young adults. The mechanism is believed to be the thickened mucus lining in the airways which apparently is protective. Also it has been shown that smokers, as a group, have elevated levels of glutathione in their epithelial linings fluids. As stated earlier, glutathione is an antioxidant that defends the lung from oxidative insults. One study compared the response to ozone in a group of smokers before and after they participated in a smoking cessation program (Emmons and Foster, 1991). They recruited 18 subjects who were habitual smokers and exposed them to 0.4 ppm ozone for 2 hours. The subjects did not show a significant ozone-induced effect on either $FEV_1$ or mid-maximal flow (MMF) rate. Six months later, 9 of the subjects had completed a smoking cessation program. Upon re-exposure to the 0.4 ppm ozone under the same conditions, the subjects showed a significant reduction in MMF. The authors stated that this study suggests the hypothesis of a protective effect of mucous proliferation may be correct. However this study must be interpreted with caution since although $FEV_1$ decreased more while the subjects were active smokers, the decrease was not significant.

**Gender as a Risk Factor for Sensitivity to Ozone**

There is some evidence that females are slightly more responsive to inhaled ozone than males. This disparity was first reported in 1985 in a study of six female subjects who were exposed to ozone or air while exercising on an ergocycle. They all underwent 10 separate exposure varying the ozone concentration from 0.0 to 0.4 ppm and also varying the exercise rate (workload). Significant decrements were seen in FVC and $FEV_1$. The results were compared to similar studies with male subjects, and indicated a larger decrement per ppm ozone. Later this same group followed up on their evaluation of gender as a risk factor for sensitivity to ozone. Male/female PFT comparisons are difficult since males have larger lung volume than females at the same height and age. The follow-up study compared females of varying lung volumes and found that lung volume per se was not a factor in predicting ozone sensitivity (Messineo et al, 1990).

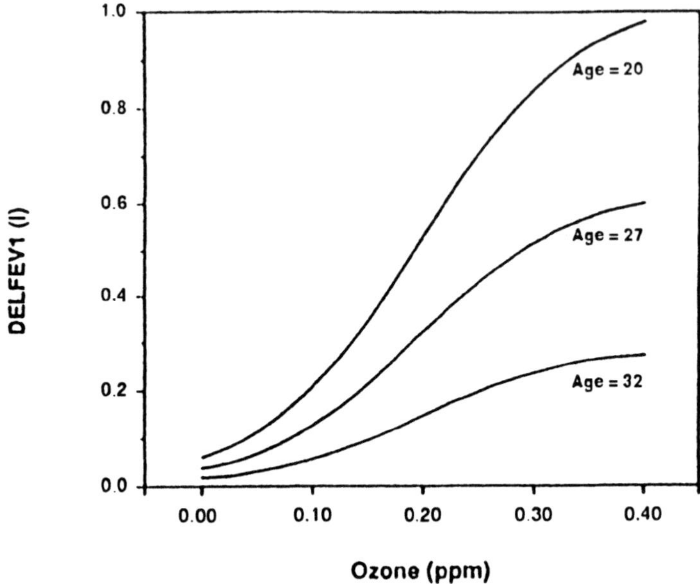


*Figure 11-3. Predicted decrements in $FEV_1$ as a function of ozone concentration for three ages. DELVE = $FEV_1$ decrement. FROM: McDonnell et al, 1993 with permission.*

**Ethnicity as a Risk Factor for Sensitivity to Ozone**

Another factor that was investigated in the search for a population susceptible to ozone was race. Only one study has been conducted and the authors concluded there was no effect of race when the analysis was conducted on the grouped data (Seal et al, 1993). However different subjects were used in different dose exposures making statistical comparisons among subjects difficult. For instance, black males had a much greater decrement in lung function after ozone than white males as seen in Table 11-4.

*Table 11-4. Mean percent change in $FEV_1$ from baseline to post ozone exposure for black and white men.*

| | % change in $FEV_1$ | |
|---|---|---|
| Ozone concentration | White men | Black men |
| 0.12 | - 2% | - 8% |
| 0.18 | - 6% | -15% |
| 0.24 | -10% | -11% |
| 0.30 | -12% | -16% |
| 0.40 | -14% | -29% |

*FROM: Adapted from Seal et al, 1993.*

However, a study comparing the response to inhaled $SO_2$ in African American subjects with asthma and Caucasian subjects with asthma found no difference based on race. There were little or no pulmonary decrements seen in either group. Both studies need to be repeated to help the understanding of potential ethnic sensitivities to certain air pollutants.

## Pharmacologic Interventions

As mentioned earlier during the summary of animal toxicology, ozone exposure also elicits a neural response in human subjects. This is the apparent mechanism for one of the most common symptoms of ozone exposure reported in human subjects in controlled laboratory exposures, pain on deep inspiration. Since deep inspiration is painful subjects do not fill their lungs to TLC and the subsequent spirometric maneuver results in a decreased FVC and $FEV_1$. This is apparently an involuntary inhibition of full inspiration. Some investigators have attempted to inhibit the inspiratory pain to verify that it is the cause for the decreased lung function tests. Lidocaine, which is a local anesthetic, was administered to young adults prior to ozone exposure (Hazucha et al, 1989). The 20% lidocaine treatment did mitigate but not totally reverse the pulmonary FVC decrements after ozone exposure. Subjects also noted less subjective pain during inspiration. An earlier study looked at the role of the parasympathetic nervous system in regulation of ozone-induced functional changes by treating subjects with atropine, a drug that mimics the parasympathic nervous system (Beckett et al, 1985). Atropine prevented the significant increase in airway resistance after ozone but only partially prevented the decrements in lung capacity measurements. This finding was interpreted as showing multiple mechanisms underlying ozone-induced lung function effects. Gong and co-workers (1988) designed a study to determine whether pretreatment with albuterol, a $beta_2$-adrenergic agonist, would block ozone-induced decrements in lung function and found that it did not. Cyclo-oxygenase inhibitors such as ibuprofen have been shown to block neutrophilic inflammation induced by ozone exposure (Hazucha et al, 1996), indicating a lessening of ozone-induced inflammation.

## Changes in Bronchial Hyperresponsiveness (BHR)

Early studies in the 1980s found increased BHR after ozone exposure (0.40 ppm) in healthy subjects as assessed by the response to inhaled methacholine. Ozone exposure has been shown to enhance the effects of exposure both to other pollutants and aeroallergens, most likely through ozone-induced BHR. On study found that prior exposure to ozone, at a sub-threshold concentration, potentiated a subsequent response to a low concentration of $SO_2$ (Koenig et al, 1990). Another study found that in a group of subjects known to be sensitive to ragweed allergen, prior exposure to ozone halved the concentration of allergen challenge needed to decrease $FEV_1$ by 20% (Molfino et al, 1991). However this study is balanced by a more recent one in subjects with mild asthma which found that a one hour exposure at rest to 0.12 ppm ozone did not enhance the bronchial response to a grass pollen challenge (Ball et al, 1996).

Several studies have examined the effect of antioxidant vitamins on the respiratory response to ozone exposure with mixed effects. One recent study examined the effects of dietary antioxidant supplementation on ozone-induced BHR in adult subjects with asthma (Trenga, 1997). This was a double-blind cross-over study, with each subject participating in each arm of the study. The supplements were 500 mg of vitamin C plus 400 IU of vitamin E versus placebos. Subjects took each regimen for 5 weeks. At the end of week 4 a subject was exposed to either air or 0.12 ppm ozone for 45 minutes during intermittent exercise. At week 5 the subject was exposed to the atmosphere not given at week 4. The sequence of the air and ozone exposures was determined randomly. A five-week wash out period separated the two regimes. Seventeen adult subjects with asthma completed the study. Assessment of the effects of the antioxidants was based on pulmonary function tests and a test for post exposure BHR that involved inhalation of $SO_2$ for two 10 minute periods while walking on a treadmill following either the ozone or air exposures. $SO_2$ is known to induce BHR in many subjects with asthma and subjects in this study were screened prior to entry for $SO_2$ sensitivity. All blood samples showed a significant treatment effect of the vitamin supplement ($p < 0.001$). The mean pre-exposure plasma vitamin C concentration was 0.9 mg/dL for placebo and 1.5 mg/dL for treatment. The mean pre-exposure vitamin E/cholesterol concentration (lipid-adjusted ratio) was 0.5 for placebo and 1.2 mg/L for treatment. $FEV_1$ and PEF values both indicated a mitigating effect of the antioxidant dietary supplements on ozone-induced BHR. PEF showed a significant within-subject difference by treatment, atmosphere and time. As seen in Figure 11-4, during vitamin treatment PEF values which dropped after ozone and the first $SO_2$ challenge, reversed after the second $SO_2$ challenge ($p = 0.009$). $FEV_1$ changes also showed a treatment effect which approached significance ($p = 0.09$). FVC also had a significant ($p = 0.03$) within-subject increase from post-exposure to post-0.1 ppm $SO_2$ challenge for the vitamin group, regardless of atmosphere. This increase indicates that the vitamin treatment may help subjects perform better after exercise and protect them from mild (0.1 ppm) $SO_2$ effects.

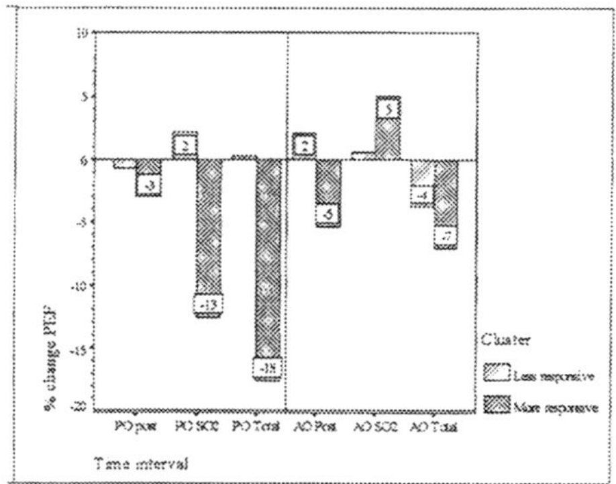


*Figure 11-4. Changes in PEF after ozone and air exposure by antioxidant treatment versus placebo. FROM: Trenga, 1997.*

Also, a study in Mexico City found that individuals treated with dietary supplements showed higher lung function values during periods of ozone pollution than those without such treatment (Romieu et al, 1998). Thus there is evidence that some of the toxic effects of ozone could be quenched by dietary antioxidants. A previous study in healthy adults reported that $O_3$-induced lipid peroxidation was attenuated by vitamin E supplementation (Dillard et al., 1978). Vitamins C and E have been observed to synergistically ameliorate ozone-induced bronchoconstriction in non-asthmatic human volunteers (Chatham et al, 1987). More recently, a field case-control study in which amateur cyclists received a combination of β-carotene, vitamin C and vitamin E reported improved post exercise lung function in the cyclists receiving supplementation (Grievink et al, 1998). The supplementation required daily intake of the vitamins for three months. A control group did not take vitamins. Lung function was measured on four occasions during the study. One drawback to this study was the use of two groups of subjects rather than a single group who experienced both regimens. Use of a repeated-measures design would have increased the statistical power of the study. Foster and co-workers (1996) examined the effects in healthy adult subjects of consecutive 3-day exposures to $O_3$ (at varying concentrations from 0.25 to 0.45 ppm) on serum α-tocopherol (vitamin E) levels. After the third day of $O_3$ exposure, serum α-tocopherol was significantly reduced compared to filtered air exposure. This again suggests that vitamin E supplementation may be beneficial to those exposed to ozone.

## Inflammatory Effects

For the first 10-15 years of intensive research investigating adverse respiratory effects of inhaled ozone, the mainstay measurement of effects was pulmonary function. Nevertheless, pulmonary function measurements were always considered to be a somewhat broad endpoint that measured only frank changes in the airways and most likely missed subtle changes in epithelial irritation or inflammation. With the development of the fiberoptic bronchoscope, it became possible to assay for some of these more subtle endpoints with the technique of bronchial alveolar lavage (BAL) which is described in Chapter 7. Koren and co-workers (1989) published an account of changes in BAL from healthy subjects after 2 hour exposure to 0.4 ppm ozone. An exposure of 0.4 ppm was associated with increases in cells known to mediate inflammation (neutrophils) and also other changes consistent with airway inflammation as shown in Table 11-5. Similar endpoints have been seen in healthy subjects after exposure to much lower concentrations of ozone (0.08 ppm).

*Table 11-5. Average changes in various endpoints of inflammation after exposure to 0.4 ppm ozone.*

| | *Air* | *Ozone* |
|---|---|---|
| *Neutrophils, %* | *1.3* | *9.9* |
| *Protein, µg/ml* | *174.9* | *340* |
| *LDH, U/ml* | *9.1* | *13.9* |
| *Fibronectin, µg/ml* | *28.5* | *53* |
| *$PGE_2$, pg/ml* | *1.3* | *3.9* |
| *IL-6, pg/ml* | *2.0* | *14.5* |

*LDH = Lactate dehydrogenase; $PGE_2$ = Prostagladins $E_2$; IL-6 = Interleukin-6.*

*FROM: Adapted from Devlin et al, 1996*

Most of the studies using BAL to evaluate the effects of air pollution on human airways have been controlled laboratory exposures. However one study was designed to test whether BAL could identify markers of inflammation in humans exposed to ozone under natural conditions. The subjects were 19 volunteer joggers who exercised in the afternoon during the summer of 1992 in Governors Island, NY (Kinney et al, 1996a). Some of the subjects were re-tested during the following winter (when ozone concentrations were low). The summertime BAL samples contained higher concentrations of markers of inflammation (Il-8, TNFα, LDH, fibronectin) than did the wintertime samples. These data suggest that recreational joggers may experience inflammatory changes in the lung when ozone concentrations range from .032- .091 ppm.

The question of whether changes in inflammatory mediators following ozone exposure are correlated with changes in pulmonary function has been addressed (Balmes et al, 1996). Twenty healthy subjects were exposed to 0.20 ppm ozone or air for 4 hr during intermittent exercise. Pulmonary function tests were performed and BAL was conducted 18 hours after the exposure. Ozone exposure increased the percent of neutrophils, total protein, and interleukin-8 for all subjects combined. However changes in $FEV_1$, Sraw, and lower respiratory symptoms were not associated with the degree of inflammation indicated by the BAL results. Thus in this study it did not appear that inflammatory responses were correlated with decrements in $FEV_1$.

# EPIDEMIOLOGICAL STUDIES

## Morbidity

Almost two decades ago, Whittemore and Korn (1980) reported evidence implicating outdoor air pollution as a factor in aggravation of asthma by showing that daily asthma attacks in asthmatics residing in Los Angeles were strongly associated with daily levels of photochemical oxidants and suspended particulate matter. Bates and Sizto (1987) found associations between ozone and hospital admissions for asthma in the summer in Southern Ontario. However these investigators found some other atmospheric species were also important. They suggested that "acid summer haze", a combination of photochemical oxidants (ozone and acid aerosols such as sulfuric or nitric acid) was the pollutant mix responsible for the effects.

Cody et al (1992) reported emergency department visits associated with ozone concentrations (and temperature) in New Jersey. Also White et al (1994) found that the average number of visits to an emergency department for asthma or reactive airway disease in children in Atlanta was 37% higher on days following elevated ozone episodes. Thurston et al (1992) found associations between ozone and hospital admissions in Buffalo, Albany, and New York City. The socioeconomic status of city appeared to be a risk factor, with stronger associations in cities with lower average socioeconomic levels. Socioeconomic status is associated with several factors known to be involved in asthma status, such as poor diet, lack of medical insurance, and crowded living conditions.

Summer camp studies of pulmonary function, symptoms and medication use are an appealing method for evaluating the real-life effects of ozone. At camp, children are involved in active outdoor events. Often the sleeping arrangements provide little protection from outdoor air, so penetration of pollutants indoors occurs. One ambient monitor can provide a quite valid exposure assessment for a group of children. Objective measurements of pulmonary function can be made daily using trained personnel. The first of these studies was conducted at a camp on the western edge of suburban New York City (Lioy et al, 1985). This first study calculated a linear regression between peak one-hour ozone concentrations for a given day and FVC, $FEV_1$, and peak flow measurements on the same day for each child. All mean slopes were negative, indicating a general tendency for decreased lung function with increasing ozone concentration. This is an important finding since children are a sensitive sub-population and the ozone concentrations were monitored close to the children's activities. Since then many other summer camp studies have been conducted. Kinney and co-workers (1996b) reported a summary of six such studies from New Jersey, southern California and Ontario, Canada. The results clearly show that children exposed under these natural conditions experience decreases in lung function.

There is one report that provides evidence that seasonal ozone exposure may cause bronchial hyperresponsiveness and also effect the immune system (Zwick et al, 1991). This study evaluated the respiratory health, allergic sensitization and immune response in 218 children from a high ozone community (Group A) and compared them with 281 children from a community with lower ozone levels (Group B). Bronchial hyperresponsiveness, measured with a methacholine challenge test, occurred more frequently in the children in Group A. Comparison of serum IgE levels showed no difference between the two groups. However there was a significantly lower helper/suppressor T cell ratio in Group A children. A description of the role of T cells in asthma is given in Ch 8. One limitation of this study is the lack of personal exposure assessment, as central monitors were used to estimate exposures.

## Mortality

Since ozone was first recognized as a major pollutant in southern California in the 1950s, epidemiologists have looked for associations between ozone exposure and mortality. Several early studies reported positive associations, however these studies may not have controlled sufficiently for meteorologic variables, autocorrelations, time trends, and other confounders. Several more recent studies have not been able to identify ozone alone as a predictor of premature mortality. Thus, at present, there is not a body of literature associating ozone exposure with daily mortality. Based on the data from animal studies and the number of morbidity outcomes associated with ozone exposure (and the known oxidant effects on the respiratory system), it is surprising that the data on ozone-mortality associations are so weak. It may be that antioxidant defenses protect people to a larger degree than expected. It may be that the appropriate epidemiology study has not yet been conducted. As pointed out throughout this book, there is always the problem of exposure assessment introducing error into tests for associations between air pollutants and health outcomes. With ozone, there is the special problem of confounding with temperature. Since ozone levels increase in direct relation to increase in ambient temperature, it is very difficult to divorce the two variables. More sophisticated analyses are beginning to control for temperature more completely. One recent study reported an association between mortality in the elderly and ozone concentrations during a hot summer in Belgium (Sartor et al, 1996). The mean number of daily deaths increased significantly for an average 6 ppb increase in ozone concentration at the same level of daily temperature. Another possibility for the lack of associations between ozone and mortality may be that the appropriate sensitive population has not yet been studied carefully. There is a report that both asthma morbidity and mortality were significantly increased during period of high ozone concentrations in New York City (Thurston et al, 1997). In general, it is not possible to use asthma mortality in the classic time series epidemiologic analysis since, with only slightly more than 5000 deaths per year in the US, deaths are not frequent enough to achieve the statistical power necessary for these tests. Examination of ozone effects on survivability such as was done for PM in the Six-City Study are not available to the best of my knowledge. In conclusion, a great deal is known about mechanisms of ozone effects but less is settled on some of the epidemiologic findings. This state of affairs is in direct contrast to the situation with PM health effects, where a vast number of epidemiologic studies point to adverse effects and yet, to date, there are very scarce reports on possible mechanisms underlying these effects.

## Chronic Effects

Another approach to understanding the relationship between ozone exposure and health effects was designed to investigate the effects of chronic exposure. Kuenzli and co-workers (1997) recruited a group of incoming college students in California and reconstructed their lifetime ozone exposure. Only subjects who had lived their entire lifetime in California were eligible. Monthly average ozone concentrations were assigned to each subject. The investigators used four time intervals (0-2; 3-5; 6-11; 12+ years) to estimate lifetime ozone exposure. They also asked about "time spent outdoors" and "time spent in moderate or heavy exercise". Two ozone metrics were used: 8 hr ozone averages (10 am to 6 pm) and hours above 0. 06 ppm. Based on these data, effective ozone exposure was then calculated over the lifetime of each student. Pulmonary function was measured with spirometry for each student (FVC, $FEV_1$, $FEF_{25-75}$, PEF, and FEF). Both $FEF_{25-75}$ and $FEF_{75}$ values were inversely associated with estimated ozone concentration. For a 0.02 ppm increase in 8-hour ozone levels, $FEF_{75\%}$ decreased 334 ml/sec (14% of the population mean) and $FEF_{25-75\%}$ decreased 420 ml/sec (7.2% of the mean). A similar correlation between estimation of lifetime ozone exposure and pulmonary function in college students was seen in a nationwide sample (Kinney et al, 1998).

A long running study of the chronic effects of air pollution, especially ozone, is the UCLA Population Study of Chronic Obstructive Respiratory Disease. The study population is drawn from census tracts in Los Angeles with either low or moderate levels of photochemical air pollution. Lung function tests have been conducted on a regular basis (several years apart) since the study's inception in 1972. In a report published in 1987 the investigators compared the rate of change in lung function dependent on pollutant category (Detels et al, 1987). In two cohorts of never-smoking residents lung function tests showed significant average decrements in the residents of more polluted community. Also, most of the spirometric tests showed significantly more rapid decline in lung function values from predicted values among adults in the more polluted community. We can expect more results from these cohorts since longitudinal studies of this type are extremely valuable.

A recent study of lung function, respiratory symptoms, activity levels, and bronchodilator use in 10-12 year old children with and without asthma is being conducted in southern California (Avol et al, 1998). The study design was such that it was expected to be an excellent opportunity to differentiate the effects of air pollution across pollution areas and between categories of children. Overall, the investigators did not find significant associations between lung function and either ozone or particulate matter. Actually, the only significant decrement in lung function associated with ozone concentrations was in healthy children. Children with asthma did appear to increase medication use in the spring but not in the summer (when ozone concentrations were highest). Thus this study raises more questions than it answers and interpretation is difficult.

## FUTURE RESEARCH NEEDS

There are always attempts to identify research needs in any scientific field. One such attempt has been under taken by the American Petroleum Institute which is developing a document on research needs to reduce uncertainly in health risk assessment for ozone. Some of the areas emphasized are 1) distinguishing between statistical significance and clinical or "causal" significance in studies, 2) improvement of exposure models, 3) use of time activity data and dietary information in estimation of exposure and identification of ozone responders, 4) study mixtures of pollutants rather than simply ozone, 5) attempt to characterize exposure error in epidemiologic studies, 6) characterize both temporal and spatial variations in ozone exposures, 7) identify and validate ozone-specific biomarkers, and 8) include information on aeroallergens in ozone field studies. Most of these concerns have been mentioned throughout the extent of this text but this list highlights the many limitations in the field of air pollution health effects.

# REFERENCES

Aris R, Christian D, Sheppard D, Balmes JR. The effects of sequential exposure to acidic fog and ozone on pulmonary function in exercising subjects. Am Rev Respir Dis 1991; 143: 85-91.

Avol EL, Linn WS, Shamoo DA, et al. Short-term respiratory effects of photochemical oxidant exposure in exercising children. J Air Poll Control Assoc 1987; 37: 158-162.

Avol EL, Nacidi WC, Rappaport EB, Peters JM. Acute effects of ambient ozone on asthmatic, wheezy, and healthy children. Report No. 82, Health Effects Instititute, 1998.

Ball BA, Folinsbee LJ, Peden DB, Kerhl HR. Allergen bronchoprovocation of patients with mild allergic asthma after ozone exposure. J Allergy Clin Imunol 1996; 98: 563-572.

Balmes JR, Chen LL, Scannell C, et al. Ozone-induced decrements in $FEV_1$ and FVC do not correlate with measures of inflammation. Am J Respir Crit Care Med 1996: 153: 9-4-909.

Bates DV, Sizto R. Air pollution and hospital admissions in Southern Ontario: The acid summer haze effect. Environ Res 1987; 43: 317-331.

Beckett WS, McDonnell WF, Horstman DH, House DE. Role of the parasympathetic nervous system in acute lung response to ozone. J Appl Physiol 1985; 59: 1879-1885.

Britton JR, Pavord ID, Richards KA, et al. Dietary antioxidant vitamin intake and lung function in the general population. Am J Respir Crit Care Med 1995; 151: 1383-1387.

Bylin G, Cotgreave, Gustafsson, et al. Health risk evaluation of ozone. Scand J Work Environ Health 1996; 22 (suppl 3): 5-104.

Calderon-Garciduenas L, Osormo-Velazquez A, et al. Histopathologic changes of the nasal mucosa in southwest metropolitan Mexico City inhabitants. Am J Pathol 1992: 140: 225-232.

Calderon-Garciduenas L, Roy-Ocotla G. Nasal cytology in southwest metropolitan Mexico City inhabitants: A pilot intervention study. Environ Health Perspect 1993; 101: 138-144.

Chatham MD, Epplker JH, Sauder LR, Green D, Kulle TJ. Evaluation of the effects of vitamin C on ozone-induced bronchoconstriction in normal subjects. Ann NY Acad Sci 1987; 498:269-79.

Chitano P, Hosselet JJ, Mapp CE, Fabbri LM. Effect of oxidant air pollutants on the respiratory system: insights from experimental animal research. Eur Respir J 1995: 8: 1357-1371.

Cody RP, Weisel CP, Birnbaum G, Lioy PJ. The effect of ozone associated with summertime photochemical smog on the frequency of asthma visits to hospital emergency departments. Environ Res 1992; 58: 184-194.

Cross CE, van der Vilet A, O'Neill CA, et al. Oxidants, antioxidants, and respiratory tract lining fluids. Environ Health Perspect 1994; 102 (suppl 10): 185-191.

Detels R, Tashkin DP, Sayre JW. Et al. The UCLA population studies of chronic obstructive respiratory disease. 9. Lung function changes associated with chronic exposure to photochemical oxidants; a cohort study among never-smokers. Chest 1987; 92: 594-603.

Devlin RB, McDonnell WF, Becket S, et al. Time-dependent changes of inflammatory mediators in the lungs of humans exposed to 0.4 ppm ozone for 2 hr: A comparison of mediators found in bronchoalveolar lavage fluid 1 and 18 hr after exposure. Toxicol Appl Pharmacol 1996; 138: 176-185.

Dillard CJ, Litov RE, Savin WM, et al. Effects of exercise, vitamin E, and ozone on pulmonary function and lipid peroxidation. J Appl Physiol 1978; 45: 927-932.

Drechsler-Parks D, Bedi JF, Horvath SM. Pulmonary function responses of older men and women to ozone exposure. Exp Gerontol1987; 22: 91-101.

Emmons K, Foster WM. Smoking cessation and acute airway response to ozone. Arch Environ Health 1991; 46: 288-295.

EPA. Air quality criteria for ozone and related photochemical oxidants. EPA/600/P-93/004. July, 1996.

Foster WM, Wills-Karp P, Tankersley CG, et al. Bloodborne markers in humans during mutliday exposure to ozone. J Appl Physiol 1996; 81: 794-800.

Gerrity TR, Weaver RA, Berntsen J, et al. Extrathoracic and intrathoracic removal of $O_3$ in tidal-breathing humans. J Appl Physiol 1988; 65: 393-400.

Gerrity TR, McDonnell WF, House DE. The relationship between delivered ozone dose and functional responses in humans. Toxicol Appl Pharmacol 1994; 124: 275-283.

Gong H, Bedi JF, Horvath SM. Inhaled albuterol does not protect against ozone toxicity in nonasthmatic athletes. Arch Environ Health 1988; 43: 46-53.

Gong H, Shamoo D, Anderson KR, Linn WS. Responses of older men with and without chronic obstructive pulmonary disease to prolonged ozone exposure. Arch Environ Health 1997; 52: 18-25.

Graham D, Henderson F, House D. Neutrophil influx measured in nasal lavages of humans exposed to ozone. Arch Env Health 1988; 43: 228-233.

Grievink L, Jansen SMA, van't Veer PV, Brunekreef B. Acute effects of ozone on pulmonary function of cyclists receiving antioxidant supplements. Occup Environ Med 1998; 55: 13-17.

Hackney JD, Linn WS, Shamoo DA, Avol EL. Responses of selected reactive and nonreactive volunteers to ozone exposure in high and how pollution seasons. In Atmospheric Ozone Research and Its Policy Implications. T. Schneider et al.(eds). Elsevier, Amsterdam, The Netherlands 1989. pp 311-318.

Hatch GE. Asthma, inhaled oxidants, and dietary antioxidants. Am J Clin Nutr 1995; 6 (suppl): 625S-630S.

Hazucha MJ, Bates DV, Bromberg PA. Mechanism of action of ozone on the human lung. J Appl Physiol 1989; 67: 1535-1541.

Hazucha MJ, Madden M, Pape G, et al. Effects of cyclo-oxgenase inhibition on ozone-induced respiratory inflammation and lung function changes. Eur J Appl Physiol 1996; 73: 17-27.

Horstman DH, Folinsbee LJ. Ozone concentration and pulmonary response relationships for 6.6 hour exposures with five hours of moderate exercise to 0.08, 0.10, and 0.12 ppm. Am Rev Respir Dis 1990; 142: 1158-1163.

Kinney PL, Nilsen DM, Lippmann M, et al. Biomarkers of lung inflammation in recreational joggers exposed to ozone. Am J Respir Crit Care Med 1996a; 154: 1430-1435.

Kinney PL, Thurston GD, Raizenne M. The effects of ambient ozone on lung function in children: a reanalysis of six summer camp studies. Environ Health Perspect 1996b; 104: 170-175

Kinney PL, Aggarwal M, Nikiforox SV, Nadas A. Methods development for epidemiologic investigations of the health effects of prolonged ozone exposure. HEI Research Report No 81. Health Effects Institute, Cambridge MA, 1998.

Koenig JQ, Covert DS, Hanley Q, et al. Prior exposure to ozone potentiates subsequent response to sulfur dioxide in adolescent asthmatic subjects. Am Rev Respir Dis 1990; 141: 377-380.

Koenig JQ. Effect of ozone on respiratory responses in subjects with asthma. Environ Health Perspect 1995; 103(suppl 2): 103-105.

Koenig JQ. The role of air pollutants in adolescent asthma. Immunol Allergy Clinics N Am 1998; 18: 61-74.

Koren HS, Devlin RB, Graham DE, et al. Ozone-induced inflammation in the lower airways of human subjects. Am Rev Respir Dis 1989; 139: 407-415.

Korrick SA, Neas LM, Dockery DW, et al. Effects of ozone and other pollutants on the pulmonary function of adult hikers Environ Health Perspect 1998; 105: 93-99.

Kuenzli N, Lurmann F, Segal M, et al. Association between lifetime ambient ozone exposure and pulmonary function in college freshmen--results of a pilot study. Environ Res 1997; 72: 8-23.

Lioy PJ, Vollmuth TA, Lippmann M. Persistence of peak flow decrement in children following ozone exposures exceeding the national ambient air quality standard. J Air Polu Control Assoc 1985; 35: 1068-1071.

Lippmann M. Ozone. In Lippmann M (ed) Environmental toxicants: Human exposures and their health effects. Van Nostrand Reinhold, New York, 1992. pp 465-519.

McBride DE, Koenig JQ, Luchtel DL, et al. Inflammatory effects of ozone in the upper airways of subjects with asthma. Am Rev Respir Crit Care Med 1994; 149: 1192-1197.

McDonnell WF, Forstman DH, Hazucha MJ, et al. Pulmonary effects of ozone exposure during exercise; dose-response characteristics. J Appl Physiol 1983; 54: 1345-1352.

McDonnell WF III, Chapman RS, Leigh MW, et al. Respiratory responses of vigorously exercising children to 0.12 ppm ozone exposure. Am Rev Respir Dis 1985; 132: 875-879.

McDonnell WF, Muller KE, Bromberg PA, Shy CA. Predictors of individual differences in acute response to ozone exposure. Am Rev Respir Dis 1993; 147: 818-825.

Messineo TD, Adams WC. Ozone inhalation effects in females varying widely in lung size: comparison with males. J Appl Physiol 1990; 69: 96-103.

Molfino NA, Wright SC, Katz I, et al . Effect of low concentrations of ozone on inhaled allergen responses in asthmatic subjects. Lancet 1991; 338: 199-203.

Mudway IS, Housley D, Eccles R, et al. Differential depletion of human respiratory tract antioxidants in response to ozone challenge. Free Rad Res 1996; 25: 499-513.

Pryor WA, Church DF. Aldehydes, hydrogen peroxide, and organic radicals as mediators of ozone toxicity. Free Rad Biol Med 1991; 11:41-46.

Pryor WA. Can vitamin E protect humans against the pathological effects of ozone in smog? Am J Clin Nutr 1991; 53:702-722.

Pryor WA. How far does ozone penetrate into the pulmonary air/tissue boundary before it reacts? Free Rad Biol Med 1992; 12:83-88.

Romieu I, Meneses F, Ramirez M, et al. Antioxidant supplementation and respiratory functions among workers exposed to high levels of ozone. Am J Respir Crit Care Med 1998; 158: 226-232.

Sartor F, Demuth C, Snacken R, Walckiers D. Mortality in the elderly and ambient ozone concentration during the hot summer, 1994, in Belgium. Environ Res 1997; 72: 109-117.

Seal E Jr, McDonnell WF, House DE, et al. The pulmonary response of white and black adults to six concentrations of ozone. Am Rev Respir Dis 1993; 147: 804-810.

Silverman F. Asthma and respiratory irritants (ozone). Environ Health Perspect 29: 131-136, 1979.

Thurston GD, Ito K, Kinney PL, Lippmann M. A multi-year study of air pollution and respiratory hospital admissions in three New York state metropolitan areas: Results for 1988 and 1989 summers. J Exp Anal Environ Epidemiol 1992; 2: 429-450.

Thurston GD, Gwynn RC. Ozone and asthma mortality/hospital admissions in New York City. Am J Respir Crit Care Med 1997; 155: 426.

Trenga CA. Dietary antioxidants and ozone-induced bronchial hyperresponsiveness in adult asthmatics. PhD Thesis. Department of Environmental Health, University of Washington, March, 1997.

Troisi RJ, Willett WC, Weiss ST, et al. A prospective study of diet and adult-onset asthma. Am J Respir Crit Care Med 1995; 151:1401-8.

White MC, Etzel RA, Wilcox WD, Lloyd C. Exacerbations of childhood asthma and ozone pollution in Atlanta. Environ Res 1994; 65: 56-68.

Whittemore AS, Korn EL. Asthma and air pollution in the Los Angeles area. Am J Public Health 70: 687-696, 1980.

Zwick H, Popp W, Wagner C, et al. Effects of ozone on the respiratory health, allergic sensitization and cellular immune system in children. Am Rev Respir Dis 1991; 144: 1075-1079.

# CHAPTER 12.

# HEALTH EFFECTS OF NITROGEN DIOXIDE

> "Given the patterns of exposure to $NO_2$ in schools and homes in some regions of Australia, we suggest that $NO_2$ exposure be considered as a determinant of respiratory illness in Australian children with persistent and unexplained symptoms or unstable asthma. Although further research is needed to prove the association, prudence about $NO_2$ exposure is warranted." (Pilotto et al, 1997).

As mentioned earlier nitrogen dioxide ($NO_2$), in outdoor air, is a primary pollutant emitted from a variety of combustion sources. These sources mainly are combustion of fossil fuels such as oil burners, automobiles, diesel trucks and buses. There are two specific issues regarding the US NAAQS for nitrogen dioxide are important to mention. One is that $NO_2$ is the only criteria pollutant that has only an annual NAAQS. All the other pollutants have shorter averaging times for the standard. An annual standard probably does not protect public health to the same extent as hourly or 24 hour standards. The lack of a shorter term standard reflects EPA's judgment that the scientific data base does not support one. However there are numerous studies documenting associations between $NO_2$ exposure and adverse respiratory effects as will be seen in this chapter. A consequence of the annual standard is the relative sparseness of $NO_2$ monitoring sites since most communities are in compliance. In fact, in Seattle, WA no $NO_2$ monitoring occurred between 1987 and 1995. Due to the lack of daily $NO_2$ data in the US, there are few recent epidemiologic studies of time series between health and air pollution that have an opportunity to examine $NO_2$ as one of the pollutants. It is kind of a viscous circle-- since there a few monitoring data there are no epidemiologic studies and since there are no epidemiologic studies there is not a data base to convince EPA to set a shorter term NAAQS. However, there are now appearing a substantial number of studies from Europe, where more extensive community monitoring for $NO_2$ has been present for years, showing associations between $NO_2$ exposures and health outcomes. Even though the US EPA has not set short-term health based standards for $NO_2$, the World Health

Organization (WHO) has set a guideline of 213 ppb $NO_2$ for a 1-hr average. The second feature that sets $NO_2$ apart from the other criteria pollutants is that the majority of documented health effects for $NO_2$ in the US are as an indoor rather than an outdoor pollutant. The sources of indoor $NO_2$ are the same as outdoors, that is, combustion. The primary source of indoor combustion releasing $NO_2$ is gas cooking stoves as will be discussed in Ch 15 on indoor air pollution.

# DEPOSITION/UPTAKE

Nitrogen dioxide is less water soluble than $SO_2$ but not as water insoluble as ozone. Thus we would expect more $NO_2$ to be taken up in the nasal passages than ozone. When inhaled, $NO_2$ is likely absorbed and converted to nitrite or nitrate in the blood. There is evidence from electron microscopy observations that $NO_2$ exposure at 2000 ppb in healthy subjects is associated with pathological changes in ciliated epithelium (Carson et al, 1993).

## CONTROLLED HUMAN STUDIES

### UPPER AIRWAY EFFECTS

In vitro studies of human nasal epithelium have provided evidence that exposure to 400 and 800 ppb $NO_2$ is associated with the synthesis of various cytokines known to be active in cellular inflammation (Devalia et al, 1993). In an in vivo laboratory study, eight subjects with seasonal allergic rhinitis were exposed to 400 ppb $NO_2$ or air in a double blind, cross over design (Wang et al, 1995). The subjects were tested out of the pollen season so that they were asymptomatic. Assessment of effects used nasal lavage and nasal airway resistance. Although neither exposure altered nasal airway resistance or levels of inflammatory cytokines, allergen challenge to mixed grass pollen after $NO_2$ induced significant signs of inflammation not seen with allergen challenge following the air exposure. Eosinophilic cationic protein (ECP-- a precursor of eosinophils that are markers of asthma aggravation) was significantly increased following $NO_2$ compared to air exposures ($p < 0.05$). This experimental design mimics a situation that could occur in the real world when sensitive subjects are exposed to $NO_2$ during the grass pollen season.

### LOWER AIRWAY EFFECTS

## Pulmonary Function Effects

Controlled laboratory studies have not found consistent, significant decrements in pulmonary function following $NO_2$ exposures at realistic ambient concentrations. One study compared the pulmonary response of adolescent subjects with and without asthma to exposures to 0.18 and 0.30 ppm $NO_2$. Neither group showed a response following the $NO_2$ exposures that was significantly different from air, however the exposures were for only one hour. Another study evaluated the response to the same concentrations of $NO_2$ in a controlled laboratory setting in a group of college athletes who were exposed while running on a treadmill for 30 minutes (Kim et al, 1991). The average minute ventilation during exposure in the group was 70 l/min and the mean heart rate was 147 beats/min indicating vigorous exercise. In this study, no statistically significant changes were observed in any pulmonary functions measured in spite of the increased ventilatory rate.

## BHR

Some investigators have shown that exposure to near ambient concentrations of $NO_2$ causes an increase in bronchial hyperresponsiveness (BHR) as measured by histamine, methacholine or carbochol challenges. A meta analysis of 20 studies using subjects with asthma and 5 studies using healthy subjects found a small but significant trend of increased BHR following $NO_2$ exposures in subjects with asthma (Folinsbee, 1992). On average, in the subjects with asthma, there was a 60% increase in BHR after $NO_2$ exposure. Evidence that $NO_2$ is associated with increased BHR also comes from studies that examined the effect of $NO_2$ exposures on a subsequent allergen challenge. These studies, by investigators in the UK, have reported that exposure to 0.4 ppm $NO_2$ for 6 hours increased the airway responsiveness to inhaled dust mite allergen in subjects with asthma (Rusznak et al, 1996). They found similar increased BHR after ozone and after exposure to $NO_2$ combined with $SO_2$. In this study, the pollutant-induced BHR lasted for 24-48 hours.

## Inflammation

$NO_2$ has been associated with inflammation of the bronchial airways as demonstrated by a study of the time course of inflammation after exposure to 2000 ppb $NO_2$ or air for 4 hours (Blomberg et al, 1997). The subjects were healthy adults. Measures of inflammation were assessed using BAL and also bronchial biopsies. $NO_2$ was associated with a 1.5-fold increase in interleukin-8 (a precursor to inflammation) and a 2.5-fold increase in neutrophils (cells present during inflammation). Jorres and associates compared the BAL response after $NO_2$ exposure in normal subjects and subjects with asthma (1995). Subjects with asthma had a small mean drop in $FEV_1$ and higher levels of inflammatory cells than seen in the normal subjects. One study compared the bronchoalveolar inflammatory effects of $NO_2$ in smokers and nonsmokers (Helleday et al, 1994). Both groups showed signs of injury after 3.5 ppm $NO_2$ although in smokers the number of macrophages increased while the numbers decreased in nonsmokers. Although this study is interesting from a mechanistic point of view, the high concentration of $NO_2$ prevents any extrapolation of this effect to individuals receiving ambient exposures.

Another group of investigators designed a study to determine whether exposure at peak indoor levels caused lower airway inflammation or impairment of host defense systems (Azadniv et al, 1998). The exposure was 2 ppm for 6 hrs. The Azadniv study evaluated BAL endpoints both at 18 hours after $NO_2$ or air exposure and, in a different set of subjects, immediately after the 6 hour exposure. The endpoints of inflammation assayed were neutrophils, and the percentages of CD4 and CD8 lymphocytes. (CD4 and CD8 cells are involved in T cell immune reactions.) Measurements of BAL cells 18 hours after exposure showed an increase in neutrophils from 2.2% to 3.1% that was significant. Also there was a small decrease in the percentage of CD8 lymphocytes. No significant changes were seen in the assays conducted on BAL cells collected immediately after exposure. These findings, using lower respiratory tract endpoints, are similar to the enhanced effects of $NO_2$ seen in the upper airways.

## PHARMACOLOGIC/ANTIOXIDANT INTERVENTIONS

It has been shown that a corticosteroid medication (fluticasone propionate nasal spray) can mitigate the BHR seen after $NO_2$ exposure (Devalia et al, 1996). That randomized study treated subjects with fluticasone or placebo for 4 weeks prior to the $NO_2$ exposure. Eosinophil cationic protein (ECP) was measured and the levels were significantly lower during the active treatment phase as compared with the placebo. Treatment with vitamin C also has been shown to mitigate the effects of inhaled $NO_2$. Mohsenin (1987) conducted a double-blind randomized study of the effect of vitamin C on $NO_2$-induced bronchial hyperresponsiveness (BHR). The subjects were 11 healthy adults with no history of asthma, rhinitis, acute respiratory illness in the past six weeks, use of vitamins, or history of cigarette smoking. The dose of vitamin C was 500 mg taken 4 times a day for three days. Subjects were exposed to 2.0 ppm $NO_2$ for one hour. Pulmonary function tests, including specific airway conductance (SGaw), were measured before and after air and $NO_2$ exposures with either placebo or vitamin C. Methacholine challenges were performed after all exposures. $NO_2$ exposure during placebo treatment induced BHR at a significantly lower concentration of methacholine than was seen during vitamin treatment (see Figure 12-1). The author concluded that the vitamin treatment completely prevented $NO_2$-induced BHR. It is interesting to compare the results of the Mohsenin study to the studies of the effects of anti-oxidant supplementation on ozone-induced BHR described in Ch 11 in subjects with asthma, although the dose of vitamin C was 4 times higher in the $NO_2$ study. Since vitamin C is water soluble and rapidly excreted from the body, the tissue levels between the two studies may not have been significantly different. The general conclusion from these studies is that antioxidant vitamins mitigate the adverse effects of inhaled oxidant gases such as $NO_2$ and ozone.

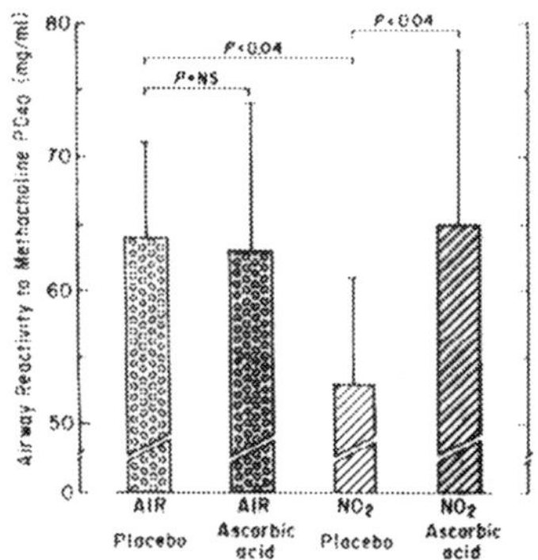


*Figure 12-1. Airway reactivity to methacholine aerosol after each exposure. Bars are means ± Standard Error. FROM: Mohsenin, 1987 with permission.*

## INFECTIVITY

Animal studies have clearly shown that $NO_2$ exposure is associated with increased risk of infection in controlled studies, including blunting respiratory defense mechanisms (mucociliary clearance and alveolar macrophages) [EPA, 1991]. When a bacterial or viral challenge is given to mice or rats following $NO_2$ exposure the rate of infectivity is greatly enhanced compared to when the challenge is given following an air exposure. Respiratory infections are an important cause of morbidity and mortality. Common respiratory infections are the common cold, influenza, acute bronchitis, and pneumonia. Respiratory syncytial virus (RSV) is a common source of winter outbreaks of acute respiratory disease. It is estimated the RSV causes 90,000 hospitalizations and 4500 deaths each year in the US from lower respiratory disease among infants and young children (MMWR, 1997). Hospitalization for pneumonia ranged between 310 and 330 per 1000 persons for 1994 (US Bureau of the Census, 1997). The death rate for pneumonia for 1996 was 13.5/100,000 persons. Unlike heart disease, where deaths are decreasing due to an aggressive campaign by the health community, infectious disease mortality including respiratory infections are increasing in the US (Armstrong et al, 1999). Thus a common air pollutant which influences infectivity would have a significant effect on quality of life and health care costs in the US. One would not expect that all cases of respiratory illness were associated with $NO_2$ exposure. Designing a study to bear evidence on the issue of $NO_2$-induced infectivity is a challenge. Whether epidemiologic studies can sort out the most likely small effect of $NO_2$ on the annual burden of respiratory disease is problematic. Thus it may be that a creative study in a controlled laboratory setting can provide clues.

One set of investigators attempted to study the relationship between $NO_2$ exposure and infectivity using human subjects in a controlled laboratory setting. Goings and associates (1989) exposed college students to $NO_2$ or air followed by instillation of a viral challenge to the nasal passages. The study was carried out over three years. The concentration of $NO_2$ was 2 ppm in year 1, 3 ppm in Year 2, and 1 ppm in Year 3. Only one of the 152 subjects developed influenza-like symptoms (after a 3 ppb exposure). There was no significant difference between the two groups in terms of increase in viral shedding. Viral shedding, the recovery of the virus from the nasal passages, was used as a sign of infectivity. In Year 1 the virus was recovered from six (29%) of the subjects after the 2 ppm $NO_2$ and only one subject after air. The difference in this year was close to statistical significance ($p = 0.07$). In Year 2 the comparison of recovered virus was 14% for $NO_2$ compared to 5% for air. This difference was not significant, although the p value was not given. Infection rates were also measured as another assessment of reduced resistance to infection. In Years 1 and 2 the infection rates were similar to those in a control group. However in Year 3, those exposed to $NO_2$ were infected more frequently than were those in the control group (91% versus 71%). Thirdly, antibody responses in the serum were evaluated; again there were no significant effects. However when the antibody responses were compared after the 2 ppm exposures in Year 1 and Year 3, the responses in the third year was significantly greater ($p < 0.03$). It appears that $NO_2$ was associated with a trend toward increased infectivity, however the effects were not significant.

Another experimental design used to evaluate the effects of $NO_2$ exposure on infectivity in human subjects used BAL and exposure of the cells washed out of the lung after either air or $NO_2$ exposure. Frampton and co-workers (1989) exposed nine healthy, non-smoking adult subjects to 0.60 ppm for 3 hr. They measured PFTs and performed BAL three and one-half hours post exposure. No significant differences in pulmonary function measurements were seen between the air and $NO_2$ exposures. Alveolar macrophages (AMs), retrieved from BAL fluid, were exposed to influenza virus by incubation. The end-point measured was the ability of the AMs to inhibit infection (by counting viral plaques). The investigators also assayed for interleukin-1 as a test for the viability of the AMs. $NO_2$ appeared to inhibit the activity of macrophages. However other studies using a similar experiment design did not see the same results.

If $NO_2$ plays a role in infectivity it may be due to interactions with rhinovirus receptor (ICAM-1) on nasal epithelium. There are no published reports of measures of ICAM-1 expression after $NO_2$ exposure, however it has been demonstrated that ozone exposure of cultured human nasal epithelial cells results in an upregulation of ICAM-1 (Beck et al, 1994). Since $NO_2$ is also an oxidant gas it is likely that $NO_2$ would also affect the expression of ICAM-1. If so, a link would be established between rhinovirus activity and $NO_2$ exposure.

## SUSCEPTIBLE POPULATIONS

### Asthma

In terms of susceptible populations, individuals with asthma appear to be more at risk for exposure to $NO_2$ (as one would expect) as seen in the Jorres study discussed earlier and also the meta analysis of studies of BHR following $NO_2$ exposure (Folinsbee, 1992). Recently Strand and co-workers reported that repeated exposure to $NO_2$ enhanced the bronchial response to birch or grass allergens in subjects with mild asthma (Strand et al, 1998). This study was carried out in Sweden where the prevalence of asthma is 5-9% in school children and 4-9% in adults, which represents a considerable increase over the last decade. Whether air pollution exposure is in any way involved in the increasing prevalence of asthma in that country is unknown.

### COPD

When compared to normal individuals of the same age, those with COPD showed more respiratory symptoms and decrements in lung function after exposure to 0.3 ppm $NO_2$ for 4 hours (Morrow et al, 1992). One striking result of this research was the range of individual responses to $NO_2$ among the subjects with COPD. One explanation could be that differences in severity of disease were responsible for the pattern of response. Interestingly the $NO_2$-induced reduction in $FEV_1$ was greater in the subjects with less severe COPD. The authors also found that patients with a history of cigarette smoking showed a greater response to the $NO_2$ exposures. This finding is in contrast to the studies of ozone exposure in subjects who smoke; these studies have found that smokers are less sensitive to the effects of ozone. Thus there is some evidence that individuals with COPD are somewhat more at risk for adverse effects from $NO_2$ exposure.

## Ice Skaters

Other individuals are susceptible due more to their exposure than to their host factors. Ice skaters exposed to $NO_2$ in skating rinks are one population at risk for adverse effects from inhalation of $NO_2$. We should take a moment to reflect why ice skaters are singled out for concern regarding $NO_2$ exposure. Ice skaters are at risk for exposure to both $NO_2$ and carbon monoxide due to the use of diesel powered equipment to smooth the surface of the ice in the skating rinks. The pairing of the release of diesel fumes which contain $NO_2$ and the cold air inversions in rinks leads to significant concentrations of $NO_2$ trapped directly in the breathing zone of the skaters. Active skaters also are breathing heavily and thus increasing their exposure. An international survey of $NO_2$ levels inside indoor ice skating rinks was conducted several years ago (Brauer et al, 1997). One-week average $NO_2$ measurements were made inside and outside skating rinks. There was a very wide range of 2-week integrated concentrations measured from 1-2680 ppb at the breathing zone of skaters and from 1- 3175 ppb in the spectator areas. In 95% of the sets of measurements, indoor $NO_2$ levels were higher than outdoor levels, indicating an indoor source of the gas. The investigators estimated the 1 hour average $NO_2$ levels from the two week samples; the data indicate that approximately 40% of the measurements exceeded the WHO guideline (213 ppb $NO_2$ for a 1-hour average). There are many anecdotal reports of individuals with asthma suffering increased symptoms while ice skating. Also there is one report of two cases of toxic pneumonitis, with delayed onset, believed to be due to exposure to $NO_2$ in a skating rink (Karlson-Stiber et al, 1996).

There also are personal $NO_2$ exposure data using passive samplers with school children (Berglund et al, 1994). This study found median daily $NO_2$ exposure levels higher in urban areas (7 ppb) than rural areas (4 ppb). The study found the highest daily values in indoor ice skating rinks (4240 ppb). Another study found $NO_2$ personal exposure in school children to range from 6 to 137 ppb (7 day averages) Linaker et al, 1996).

# EPIDEMIOLOGY

## Morbidity

There has been a suspicion that $NO_2$ was negatively associated with respiratory morbidity for years, partially based on the suggested relationship between $NO_2$ and infectivity described above. One of the earliest studies of acute respiratory illness was conducted in Chattanooga school children (Shy et al, 1971). The study compared illness rates in areas with high or low concentrations of $NO_2$. Respiratory illness rates were consistently higher in all family members in the high $NO_2$ area over the entire 24 week period of study. A relative excess of 19% was found among the exposed families as compared to families in the lower $NO_2$ area.

Much of the epidemiology of the health effects of $NO_2$ has been centered on childhood respiratory disease associated with the presence of a gas cooking stove in the home. Children living in homes with gas cooking stoves are considered to be an "at risk" population due to their increased exposure. A gas cooking stove can emit a high concentration of $NO_2$ when first lighted. Figure 12-2 shows peak exposures when a burner and the oven of a gas stove are ignited. Although these concentrations seen in Figure 12-2 are at the high end of levels found in outdoor air, it is not as high as peak levels found in the proximity of gas-cooking stoves. Those levels are as high as 1843 $\mu g/m^3$ (nearly 1000 ppb) for a 3 minute average (EPA, 1991). The population in the US potentially exposed to these higher indoor concentrations of $NO_2$ far exceeds the population potentially exposed outdoors to elevated concentrations of $NO_2$.

At this point, the literature of exposures to $NO_2$ from gas cooking stoves and respiratory disease will be summarized. One study found decreases in forced expiratory volume in one second ($FEV_1$) and forced vital capacity (FVC) in subjects with asthma when $NO_2$ concentrations were greater than 300 ppb (Goldstein et al, 1987). As part of the Harvard Six-City study Neas and co-workers (1991) reported, in a sample of 1500+ children aged 7-11 years, that annual exposure to $NO_2$ was positively associated with increased respiratory symptoms. The symptoms recorded were shortness of breath with wheeze, chronic wheeze, chronic cough, and chronic phlegm production. In general, a 15 ppb increase in the annual $NO_2$ levels within households was significantly associated with the cumulative incidence of these lower respiratory tract symptoms. The overall odds ratio was 1.4 (95% CI, 1.1-1.7). Girls had a stronger association than boys (odds ratio of 1.7 vs 1.2). There was no association between presence of a gas stove in the home and standard pulmonary function measurements. These results could be interpreted as showing an adverse respiratory effect from $NO_2$ at concentrations found indoors. A recent study exploring childhood asthma and indoor environmental risk factors followed a subset of the 457 children with $NO_2$ badges and found the odds ratios for asthma aggravation increased with each level (0 to ≤ 10 ppb, > 10 to ≤ 15 ppb, and > 15 ppb) of $NO_2$ (Infante-Rivard, 1993). However, one large study found no increase in respiratory infection of infants living in gas vs electric range homes (Samet et al, 1993). A meta-analysis of many of these studies found that, although the individual studies were inconclusive, the merged analysis yielded an odds ratio of 1.18 (CI 1.11-1.25) for increased respiratory illness associated with long-term exposure to 30 $\mu g/m^3$ $NO_2$. (~ 15 ppb) (Hasselblad et al, 1992). The lack of detailed exposure information may have decreased the ability of the studies to detect a positive relationship. One study in Stockholm, conducted since the meta-analysis, found that the risk of wheezing bronchitis in girls aged 4 months to 4 years was related to a time-weighted time of outdoor $NO_2$ exposure. A gas stove in the home appeared to be the major source of $NO_2$ (Pershagen et al, 1995). Although the vast majority of the studies of effects of indoor $NO_2$ exposure have targeted children, a recent study found an association between $NO_2$ exposure from gas cooking stoves and aggravation of asthma in female adults (Jarvis et al, 1996).

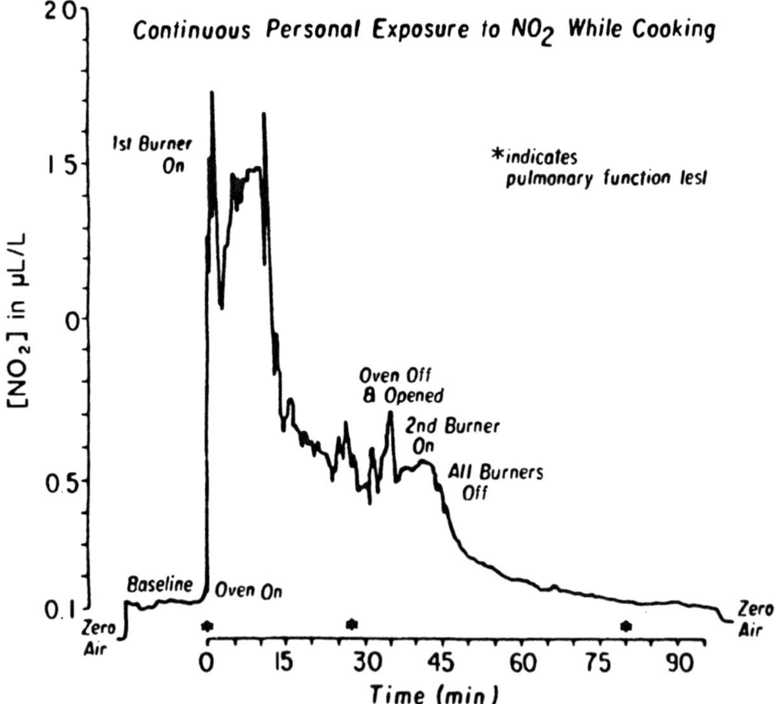


*Figure 12-2. $NO_2$ exposure to a woman while cooking. FROM: Goldstein and Andrews, 1987 with permission.*

In the Samet and Neas studies, $NO_2$ was measured indoors using Palmes tubes which integrated the concentration over two weeks and one week respectively. The Infante-Rivard study used personal $NO_2$ badges to calculate the $NO_2$ exposure. Thus these studies are greatly strengthened by the presence of adequate exposure assessment as discussed in Ch 5. Average indoor concentrations over a two-week period in homes with gas versus electric cooking stove are shown in Figure 12-3.

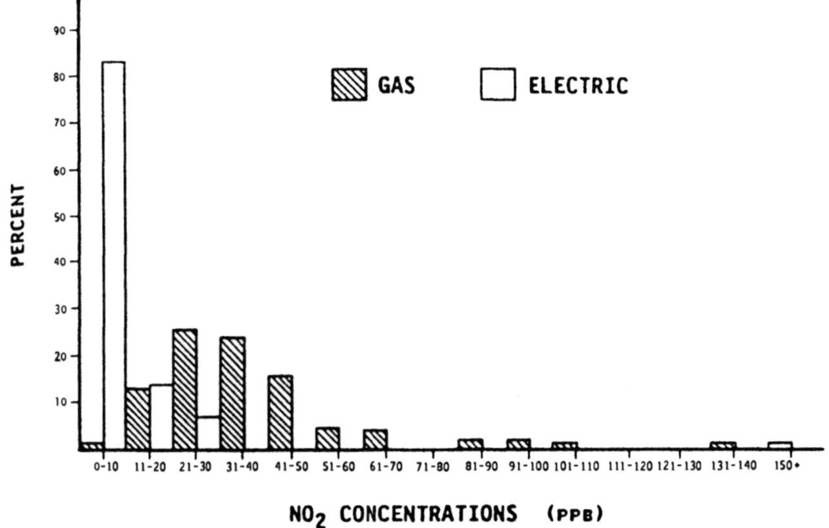


*Figure 12-3. Two-week integrated $NO_2$ concentrations in homes in Albuquerque. FROM: Samet, 1991 with permission.*

Epidemiological studies have also investigated the relationship between adverse health events and outdoor concentrations of $NO_2$. In one study the effects of ambient $NO_2$ concentrations on lung function was investigated in 423 schoolchildren (Frischer et al, 1993). Spirometry was performed over a 6-month period and linear regressions were used to estimate the effect of $NO_2$ on subsequent lung function values. The results showed that for each ppb decrease in $NO_2$ there was a 0.8 ml increase in FVC. Another European study with schoolchildren found similar results over a longer time period. Moseler and associates (1994) conducted a 2 and 1/2 year study of the influence of outdoor $NO_2$ on lung function in 467 in Germany in children with symptoms of asthma. Five measures of lung function including the percent of measured $FEV_1$ compared to predicted $FEV_1$ values showed a negative relationship with outdoor $NO_2$ levels of approximately 27 ppb or greater. One study was conducted using Danish subjects with non-allergic asthma (Moseholm et al, 1993). A strength of this study is that, since the subjects did not have allergy, the confounding from pollen and fungi exposure was greatly reduced if not removed. The study found that increased levels of both

$NO_2$ and $SO_2$ were related to decreased peak flow values. Medication use and high ambient temperature also were related to decreases in peak flow.

There is one study which has shown an association between outdoor $NO_2$ levels and asthma visits to emergency departments in Valenica, Spain using time-series analysis (Tenias et al, 1998). Poisson regression showed that an increase of 8 ppb in ambient air pollution was associated with a relative risk of an emergency visit for asthma of 1.07 (confidence intervals of 1.02 to 1.13) for a 24 hour average of $NO_2$. In this study the association was more consistent for $NO_2$ and ozone than for particulate matter and $SO_2$.

## Mortality

In the winter of 1991 London experienced a period of air stagnation and ambient air pollutant levels increased dramatically. (This is somewhat reminiscent of the famous London "killer fog" episode described in the Ch 10 on the health effects of particulate matter, although currently the source of air pollution is more likely to be diesel fueled vehicles than coal burning.) The number of deaths and hospital admissions occurring in Greater London during the week of the episode were compared to a control week (Anderson et al, 1995). When all-age mortality excluding accidental deaths was compared to deaths during the control period using a log linear model, a significant increase was seen for cardiovascular diseases. There were increases for all-age respiratory diseases as well but these did not reach statistical significance. However, for individuals over the age of 65 years, hospital admissions for all respiratory diseases were significantly increased. All air pollutants were elevated during the period however $NO_2$ levels rose to record concentrations. The authors conclude that air pollution, especially $NO_2$, is a plausible explanation for the increases in deaths and hospital admissions.

## Chronic Effects

As has been mentioned repeatedly throughout this book, most of the information on the health effects of air pollutants is derived from studies designed to interpret the short-term or acute effects of the pollutants. Obviously studies of short-term effects are logistically much easier to conduct. However in the past year or so, several investigators have begun to design experiments which are oriented at an evaluation of more longitudinal effects of pollutants. Studies of estimated lifetime exposure to particulate matter and ozone are discussed in Chs 10 and 11 respectively. Regarding the effects of longitudinal exposure to $NO_2$, a study in Switzerland looked at associations between lung function and estimated average exposure to $NO_2$ in eight areas of the country (Schindler et al, 1998). This cross sectional study recruited a random sample of 17,300 adults between the ages of 18 and 60 years of age. Of these, spirometry measurements were made successfully on 9651 subjects. Exposure was estimated from zones of residence taking into account the presence or absence of a gas cooking stove. The analysis found a negative relationship between lung function values and estimated $NO_2$ exposure. Each 8 ppb increase in $NO_2$ was associated with an estimated 0.74% decrease in $FEV_1$. The association would be somewhat weaker if based solely on the outdoor $NO_2$ exposures. However taken with the other recent studies of the effects of estimated life-time exposure to air pollution and lung function, this study joins a group of studies which cause a great deal of concern for the relationship between air pollution and respiratory health. The absolute value of $FEV_1$ for an individual is known to be an extremely accurate predictor of mortality. Therefore any environmental conditions that are related to lower $FEV_1$ values in the population must be regarded very seriousl

## CONCLUSIONS

In my opinion the respiratory health effects of $NO_2$ have been under-rated. An annual increase of 15 ppb $NO_2$ in indoor concentration is associated with significant adverse respiratory effects. This concentration is considerably lower than the annual NAAQS for outdoor $NO_2$. It appears that the present standard (53 ppb averaged on an annual basis ) is not protective. Future research most likely will show that individuals with asthma will be shown to be quite sensitive--we just don't know the endpoint yet--and that the infectivity effects will be found human populations. I am not alone in this opinion. $NO_2$ is an international public health problem. An editorial used at the beginning of this chapter is reprinted here:

> "Given the patterns of exposure to $NO_2$ in schools and homes in some regions of Australia, we suggest that $NO_2$ exposure be considered as a determinant of respiratory illness in Australian children with persistent and unexplained symptoms or unstable asthma. Although further research is needed to prove the association, prudence about $NO_2$ exposure is warranted." (Pilotto et al, 1997).

# REFERENCES

Anderson HR, Limb ES, Bland JM, et al. Health effects of an air pollution episode in London, December, 1991. Thorax 1995; 50: 1188-1193.

Armstrong GL, Conn LA, Pinner RW. Trends in infectious disease mortality in the United States during the $20^{th}$ century. JAMA 1999; 281: 61-66.

Azadniv M, Utell MJ, Morrow PE, et al. Effects of nitrogen dioxide exposure on human host defense. Inhal Toxicol 1998; 10: 585-601.

Beck NB, Koenig JQ, Luchtel DL. Ozone can increase the expression of intercellular adhesion molecule-1 and the synthesis of cytokines by human nasal epithelial cells. Inhal Toxicol 1994; 6: 345-357.

Berglund M, Braback L, Bylin G, et al. Personal $NO_2$ exposure monitoring shows high exposure among ice-skating schoolchildren. Arch Environ Health 1994; 49: 17-24.

Blomberg A, Krishna MT, Bocchino V, et al. The inflammatory effects of 2 ppm $NO_2$ on the airways of healthy subjects. Am J Respir Crit Care Med 1997; 156: 418-424.

Brauer M, Lee K, Spengler JD, et al. Nitrogen dioxide in indoor ice skating facilities: An international study. J Air Waste Manage Assoc 1997; 47: 1095-1102.

Carson JL, Collier AM, Hu S-CS, Devlin RB. Effect of nitrogen dioxide on human nasal epithelium. Am J Respir Cell Mol Biol 1993; 9: 264-270.

Devalia JL, Campbell AM, Sapsford RJ, et al. Effect of nitrogen dioxide on synthesis of inflammatory cytokines expressed by human bronchial epithelial cells *In Vitro*. Am J Respir Cell Mol Biol 1993; 9: 271-278.

Devalia JL,Wang JH, Rusznak C, et al. The effect of fluticasone propionate aqueous nasal spray on allergen-induced inflammatory changes in the nasal airways of allergic rhinitics following exposure for six hours to 400 ppb nitrogen dioxide. Eur Respir J 1996: 9 (Suppl 23): 86S.

EPA. Air qualtiy criteria for oxides of nitrogen. EPA-/600/8-91/049. August, 1991.

Folinsbee LJ. Does nitrogen dioxide exposure increase airways responsiveness? Toxicol Indust Health 1992; 8: 273-283.

Frampton MJ, Smeglin AM, Roberts NJ Jr, et al. Nitrogen dioxide exposure in*Vivo* and human alveolar macrophage inactivation of influenza virus in *Vitro*. Environ Res 1989; 48: 179-192.

Frischer T, Studnicka M, Beer E, Neumann M. The effects of ambient $NO_2$ on lung function in primary schoolchildren. Environ Res 1993; 62: 179-188.

Goings SAJ, Kulle TJ, Bascom R, et al. Effect of nitrogen dioxide exposure on susceptibility to influenza A virus infection in healthy adults. Am Rev Respir Dis 1989; 139: 1075-1081.

Goldstein I, Andrews LR. Peak exposures to nitrogen dioxide and study design to detect their acute health effects. Environ Int 1987; 13: 285-291.

Hasselblad V, Eddy DM, Kotchmar DJ. Synthesis of environmental evidence: Nitrogen dioxide epidemiology studies. J Air Waste Manage Assoc 1992; 42: 662-671.

Helleday R, Sandstrom T,. Stjernberg N. Differences in bronchoalveolar cell response to nitrogen dioxide exposure between smokers and nonsmokers. Eur Respir J 1994; 7: 1213-1220.

Infante-Rivard C. Childhood asthma and indoor environmental risk factors. Am J Epidemiol 1993; 137: 834-844.

Jarvis D, Chinn S, Luczynska C, Burney P. Association of respiratory symptoms and lung function in young adults with use of domestic gas appliances. Lancet 1996; 347: 426-431.

Jorres R, Nowak D, Grimminger F, et al. The effect of 1 ppm nitrogen dioxide on bronchoalveolar lavage cells and inflammatory mediators in normal and asthmatic subjects. Eur Respir J 1995; 8: 416-424.

Karlson-Stiber C, Hojer J, Sjoholm A, et al. Nitrogen dioxide pneumonitis in ice hockey players. J Intern Med 1996; 239: 451-456.

Kim SU, Koenig JQ, Pierson WE, Hanley QS. Acute pulmonary effects of nitrogen dioxide exposure during exercise in competitive athletes. Chest 1991; 99: 815-819.

Linaker CH, Chauhan AJ, Inskip H, et al. Distribution and determinants of personal exposure to nitrogen dioxide in school children. Occup Envirion Med 1996; 53: 200-203.

MMWR. Update: Respiratory syncytial virus activity--United States, 1997-98 season. MMWR, 1997; 46: 1163-1165.

Mohsenin V. Effect of vitamin C on $NO_2$-induced airway hyperresponsiveness in normal subjects. Am Rev Respir Dis 1987; 136: 1408-1411.

Morrow PE, Utell MJ, Bauer MA, et al. Pulmonary performance of elderly normal subjects and subjects with chronic obstructive pulmonary disease exposed to 0.3 ppm nitrogen dioxide. Am Rev Respir Dis 1992; 145: 291-300.

Moseholm L, Taudorf E, Forsig A. Pulmonary function changes in asthmatics associated with low-level $SO_2$ and $NO_2$ air pollution, weather, and medicine intake. Allergy 1993; 48: 334-344.

Moseler M, Hendel-Framer A, Kaarmaus W, et al. Effect of moderate $NO_2$ air pollution on the lung function of children with asthmatic symptoms. Environ Res 1994; 67: 109-124.

Neas LM, Dockery DW, Ware JH, et al. Association of indoor nitrogen dioxide with respiratory symptoms and pulmonary function in children. Am J Epidemiol 1991; 134: 204-218.

Pershagen G, Rylfander E, Norberg S, et al. Air pollution involving nitrogen dioxide exposure and wheezing bronchitis in children. Int J Epidemiol 1995; 24: 1147-1153.

Pilotto LS, Douglas RM, Samet JM. Nitrogen dioxide, gas heating and respiratory illness. Med J Australia 1997; 167: 295-296.

Rusznak C, Devalia JL, Davies RJ. The airway response of asthmatics to inhaled allergen after exposure to pollutants. Thorax 1996; 51: 1105-1108.

Samet JM. Skipper LW, Cushing AH, et al. Nitrogen dioxide and respiratory illnesses in infants., Am J Rrespir Dis 1993; 148: 1258-1265.

Samet JM. Nitrogen dioxide. IN Indoor Air Pollution, Samet and Spengler (eds). Johns Hopkins University Press, Baltimore, 1991.

Schindler C, Ackermann-Liebrich U, Leuenberger P, et al. Associations between lung function and estimated average exposure to $NO_2$ in eight areas of Switzerland. Epidemiology 1998; 9: 405-411.

Shy CM, Creason JP, Pearlman ME, et al. The Chattanooga school children study: effects of community exposure to nitrogen dioxide. J Air Pollut Control Assoc 1971; 20: 582-588.

Strand V, Svartengen M, Rak S, et al. Repeated exposure to an ambient level of $NO_2$ enhances asthmatic response to a nonsymptomatic allergen dose. Eur Respir J 1998; 12: 6-12.

Tenias JM, Ballester F, Rivera ML. Association between hospital emergency visits for asthma and air pollution in Valencia, Spain. Occup Environ Med 1998; 55: 541-547.

US Bureau of the Census. Statistical abstracts of the United States, 1997 (117th edition). Washington, DC, 1997.

Wang JH, Devalia JL, Duddle JM, et al. Effect of six-hour exposure to nitrogen dioxide on early-phase nasal response to allergen challenge in patients with a history of seasonal allergic rhinitis. J Allergy Clin Immunol 1995; 96: 669-676.

# CHAPTER 13.

# HEALTH EFFECTS OF CARBON MONOXIDE

Carbon dioxide is a tasteless, odorless, colorless, and non-irritating gas produced by incomplete combustion of organic materials. It is present wherever combustion occurs. The major outdoor sources are vehicles (automobiles, light duty truck, heavy trucks, and buses). It also is emitted from power plants and industrial sources. CO also is an indoor pollutant that can reach toxic levels when indoor combustion sources are used without adequate ventilation. The rule is: Never use any indoor combustion source indoors without a vent. This applies to wood stoves and fireplaces (always make certain the fire in completely out before closing the draft to a fireplace). It also applies to kerosene and propane space heaters and even charcoal fueled burners such as a hibachi. During the holiday season of 1998, a family was found dead in their home due to CO poisoning from use of an unvented hibachi used for warmth. Another source of CO poisoning is in vehicles without proper ventilation or with malfunctioning mufflers. Every year there are reports of accidental deaths from children and adults who are poisoned from staying in a vehicle with the engine running and improper ventilation. The Morbidity/Mortality Weekly report from the Centers for Disease Control and Prevention reported 11,547 unintentional CO deaths in the US from 1979-1988 (MMWR, 1996). Of these 57% were caused by motor-vehicle exhaust; 83% of these deaths occurred in stationary vehicles. Deaths can occur in a vehicle in a garage even if the windows are open. CO is the leading cause of poisoning in the US. CO in the home can reach toxic levels from the use of defective or improperly installed appliances, from poor ventilation due to blocked chimneys or vents, and from idling vehicles in attached garages.

Carbon monoxide poisoning occurs as a result of tissue hypoxia (lack of oxygen). CO, when inhaled, can diffuse rapidly across the alveolar membrane. It binds reversibly to hemoglobin with a 200-fold greater affinity than oxygen. Symptoms of CO poisoning are confusion, nausea, headache, dizziness, fatigue, drowsiness, and coma sometimes resulting in death. Unfortunately many of these symptoms are non-specific and CO poisoning is frequently overlooked. The primary target organs for CO poisoning are those tissues in the body which use the majority of systemic oxygen; that is, the central nervous system and the heart.

Note that the respiratory system is not considered to be a target organ for CO adverse effects. Since the mechanism of CO effects appears well known, there is not an extensive data base of animal toxicology, controlled human studies, or epidemiology of air pollution studies centered on CO. With the recent epidemiologic report associating PM with both respiratory and cardiovascular mortality, there has been an increase in the number of epidemiologic studies evaluating the adverse effects of CO. The literature from controlled laboratory studies of CO effects is mainly studies of the cardiovascular effects of CO. The NAAQS for CO is based on its effects on the heart.

## EXPOSURE ASSESSMENT

In the 1970s and 1980s many US cities were out of compliance with either the 1 hour or the 8 hour NAAQS for CO. However with more stringent controls on automobiles, CO levels have dropped remarkably in the past decade (see Figure 4-1). Only 12 million individuals lived in areas not meeting the current CO standard in 1995. It may be that some personal exposures are still high dependent on personal activity patterns. One study was conducted to evaluate personal exposures to CO in Athens, Greece (Vellpoulou and Ashmore,1998). The objective of the study was to evaluate CO exposures in commuters. Personal CO samplers (badges) were used on a total of 50 subjects during the winter of 1992 and the spring of 1993. That study found higher exposures to CO while persons were commuting than while at home or in offices. Mean CO concentrations in various microenvironments are shown in Table 13-1.

*Table 13-1. Mean CO concentrations ($mg/m^3$) in different situations.*

| | *Sample* | *Mean* | *Min* | *Max* |
|---|---|---|---|---|
| *City Pavement* | *9* | *14.9* | *5.9* | *25.0* |
| *Garage* | *5* | *13.8* | *6.5* | *30.8* |
| *Café* | *17* | *12.1* | *5.1* | *24.4* |
| *Pub/Bar* | *6* | *9.7* | *5.8* | *11.7* |
| *Residential street* | *13* | *7.3* | *4.9* | *12.9* |

| | | | | |
|---|---|---|---|---|
| *Office (s)* | *12* | *7.1* | *5.6* | *10.0* |
| *Home (s)* | *8* | *6.4* | *4.9* | *8.7* |
| *Office (ns)* | *34* | *2.9* | *0* | *5.5* |
| *Home (ns)* | *26* | *2.1* | *0.2* | *4.6* |

*s= smoker; ns = nonsmoker*

*FROM: Adapted from Vellopoulou and Ashmore, 1998.*

It is instructive to compare the CO values in offices and homes dependent on the presence or absence of smokers. Outdoor CO exposure is determined by distance from traffic congestion and idling vehicles. In a study of background CO in Seattle, 8 hour mean concentrations of CO were 0.7 ppm in an a park, 1.62 ppm in traffic, and 2.73 ppm on a busy street.

## HEALTH EFFECTS OF CO

## CONTOLLED STUDIES

Most controlled studies of CO exposure have been conducted to examine the effects of inhaled CO on the cardiac system. The classic study design has been evaluation of low levels of CO in individuals with pre-existing cardiac disease. One typical study was carried out with 10 subjects who breathed either air, 50 ppm CO or 100 ppm CO for 4 hours during moderate exercise on a treadmill (Anderson et al, 1973). Blood was drawn to determine carboxyhemoglobin levels, an accurate biomarker for exposure to CO. Actually the presence of a verified biomarker is a unique feature of CO exposure that is not available for the other criteria pollutants. The mean duration of exercise tolerance was shortened with both 50 and 100 ppm CO in comparison with air ($p < 0.005$). Caroxyhemoglobin levels increased from 1.3% with air to 2.9% with 50 ppm and 4.5% with 100 ppm CO. A more recent study was conducted ten years ago and essentially confirmed the above results (Allerd et al, 1989). Those investigators also concluded that low levels of CO exposure during moderate exercise exacerbated myocardial ischemia. Carboxyhemoglobin levels associated with air exposures were 0.6 % with 117 ppm CO were 2.0 %, and with 253 ppm CO were 3.9 %. The Allred study assessed an adverse effect of CO by measuring the time of exposure prior to changes in the ST segment of EKG record. A ST segment change is considered indicative of myocardial ischemia. The relationship between ST-segment changes and COHb levels after exercise is shown in Figure 13-1. Other controlled studies have found an association between CO exposure and ventricular arrhythmias (Sheps et al, 1990). This is especially interesting in light of the more recent epidemiologic studies associating death and outdoor CO concentrations.

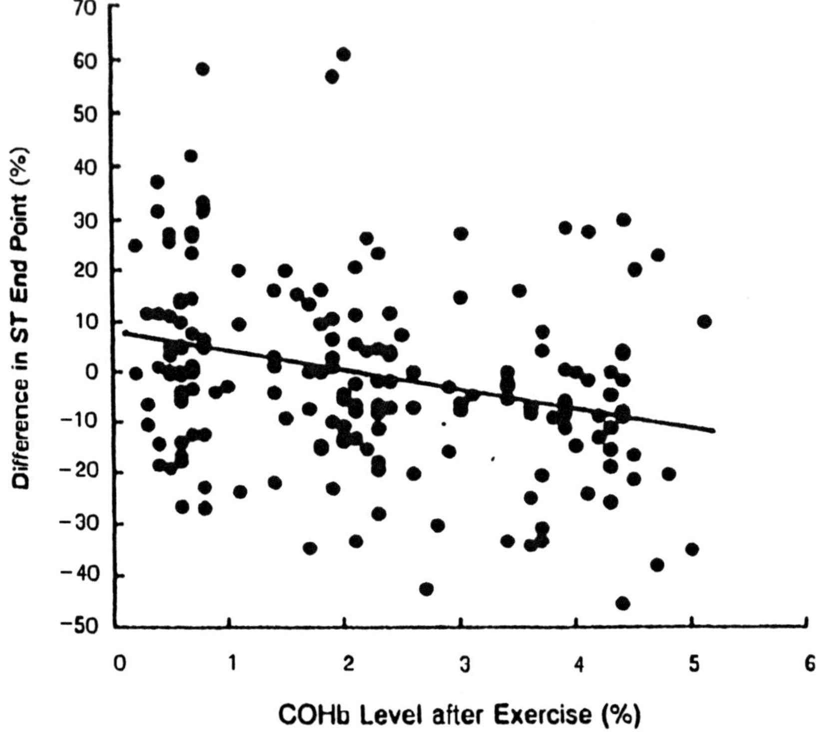


*Figure 13-1. Dose-response relation between the percentage change in the length of time to the threshold ischemic ST-segment change and the carboxyhemoglobin level after exercise. FROM: Allred et al, 1989 with permission.*

# EPIDEMIOLOGICAL STUDIES

During the 1980s and early 1990s, the majority of the epidemiologic studies looking for health effects of air pollutants were centered on PM air pollution. However in the past few years, there has been increasing interest in the effects of daily concentrations of CO on daily health outcomes such as hospital admissions or death. Some of these studies have been designed to investigate the effects of ambient CO concentration on cardiac outcomes. Others have been studies where the results show that CO concentrations are as strongly associated with health outcomes such as asthma hospital visits as PM. This has led to confusion since there is no known relationship between asthma and CO exposures. This confusing situation will be discussed later. At this point some of the epidemiologic studies examining a relationship between CO and health will be reviewed.

A study of seven large US cities found evidence for an association between hospitalization for congestive heart failure and ambient CO concentrations (Morris et al, 1995). In that study, the relative risk of hospital admission for congestive heart failure associated with a 10 ppm increase in outdoor CO ranged from 1.10 in New York city to 1.37 in Los Angeles. Relative risk of a hospital admission for congestive heart failure associated with air pollution for the seven cities is given in Table 13-2. In the US, an estimated 3 million individuals suffer from congestive heart failure. According to the primary author, even on days when CO concentrations in a given city were well below the NAAQS, a 20 to 40 percent increase in admissions was associated with CO concentrations measured on the same day.

In a recent study from Seattle, an interquartile range change in $PM_{10}$, $PM_{25}$, $PM_{10-2.5}$, and CO were all associated with daily hospital admissions for asthma in individuals under the age of 65 years (Sheppard et al, 1999). A similar result was seen in Spokane for emergency department visits for asthma. Since, as mentioned above, there is no plausible biologic association between asthma aggravation and CO, the authors concluded the CO may be acting as a surrogate for stagnant, polluted air.

Since any scientific field is a fluid, constantly changing, accumulation of data, it is possible that in the next ten years or so, we may change our understanding of the action of CO on the lung and on asthma specifically. In fact, there are some recent data suggesting a link between CO and pulmonary vascular stress (Thom et al, 1999). The objective of that study was to determine whether exposure to ambient levels of CO could cause lung damage through the mechanism of nitric oxide (NO) mediation. Lung capillary leakage was significantly increased 18 hours after rats had been exposed to 50 ppm CO. There was a concomitant increase in NO concentrations.

## CONCLUSION

At this point, a relationship between CO exposures and respiratory health outcomes is unknown. There is a long-standing relationship between CO exposure and cardiovascular health. It is very difficult to attribute health outcomes to one specific pollutant due to the complex mixture of urban air pollution. This has been a recurring theme of this book as the health effects of each criteria pollutant was discussed. Currently investigators are attempting to construct better statistical models to reduce uncertainly. The next decade promises to be a very exciting period for air pollutant and health. The present monitoring system, that places CO ambient monitors in street canyon where levels are likely to be highest, may have to be modified in order to provide sufficient data to examine the relationship between ambient CO exposures to the general population and health outcomes. Everyone agrees that the street canyon CO monitors do not represent residential exposure. Also studies of personal 24-hour exposures to CO would be useful in helping determine actual exposure and the relationship between personal and fixed site monitor values.

*Table 13-2. Relative risks of congestive heart failure admission among medicare patients as a function of ambient pollutant levels. FROM: Morris et al, 1995 with permission.*

| | Single-Pollutant Model | | Multipollutant Model | |
|---|---|---|---|---|
| Pollutant and City | Relative Risk | 95% Confidence Interval | Relative Risk | 95% Confidence Interval |
| Carbon monoxide | | | | |
| Los Angeles | 1.36 | 1.28, 1.46 | 1.39 | 1.23, 1.56 |
| Chicago | 1.29 | 1.16, 1.44 | 1.23 | 1.07, 1.43 |
| Philadelphia | 1.17 | 1.05, 1.31 | 1.22 | 1.05, 1.41 |
| New York | 1.10 | 1.03, 1.18 | 1.05 | 0.97, 1.14 |
| Detroit | 1.24 | 1.11, 1.39 | 1.38 | 1.17, 1.63 |
| Houston | 1.11 | 0.97, 1.26 | 1.25 | 1.05, 1.49 |
| Milwaukee | 1.29 | 1.07, 1.57 | 1.26 | 0.89, 1.77 |
| Nitrogen dioxide | | | | |
| Los Angeles | 1.15 | 1.10, 1.19 | 0.98 | 0.91, 1.05 |
| Chicago | 1.17 | 1.07, 1.27 | 1.06 | 0.92, 1.22 |
| Philadelphia | 1.03 | 0.95, 1.12 | 0.95 | 0.84, 1.08 |
| New York | 1.07 | 1.02, 1.13 | 1.08 | 1.01, 1.16 |
| Detroit | 1.04 | 0.92, 1.18 | 1.01 | 0.88, 1.15 |
| Houston | 0.99 | 0.88, 1.10 | 0.83 | 0.70, 1.00 |
| Milwaukee | 1.05 | 0.89, 1.23 | 0.87 | 0.68, 1.12 |
| Sulfur dioxide | | | | |
| Los Angeles | 1.60 | 1.41, 1.82 | 1.00 | 0.81, 1.24 |
| Chicago | 1.05 | 1.00, 1.10 | 1.00 | 0.95, 1.06 |
| Philadelphia | 1.01 | 0.96, 1.06 | 0.99 | 0.94, 1.05 |
| New York | 1.04 | 1.01, 1.08 | 1.01 | 0.97, 1.06 |
| Detroit | 1.00 | 0.95, 1.06 | 0.98 | 0.91, 1.06 |
| Houston | 1.07 | 0.97, 1.17 | 1.11 | 0.99, 1.24 |
| Milwaukee | 1.07 | 0.99, 1.15 | 1.07 | 0.97, 1.19 |
| Ozone | | | | |
| Los Angeles | 1.06 | 1.01, 1.11 | 1.03 | 0.97, 1.09 |
| Chicago | 1.03 | 0.93, 1.14 | 0.98 | 0.87, 1.10 |
| Philadelphia | 0.95 | 0.87, 1.05 | 0.98 | 0.89, 1.09 |
| New York | 0.89 | 0.81, 0.97 | 0.84 | 0.76, 0.92 |
| Detroit | 0.90 | 0.78, 1.05 | 0.92 | 0.76, 1.12 |
| Houston | 0.99 | 0.90, 1.08 | 0.98 | 0.87, 1.11 |
| Milwaukee | 1.00 | 1.90, 0.53 | 0.84 | 0.69, 1.02 |

*Note.* The multipollutant model included all four pollutants. All models included temperature, month, day of week, and year. Values refer to the relative risk associated with an increase of 10 ppm of carbon monoxide, 0.1 ppm of nitrogen dioxide, 0.05 ppm of sulfur dioxide, or 0.12 ppm of ozone.

## REFERENCES

Allred E, Bleeker ER, Chaitman BR, et al. Short-term effects of carbon monoxide exposure on the exercise performance of subjects with coronary artery disease. N Engl J Med 1989; 321: 1426-1432.

Anderson EW, Andelman RJ, Strauch JM, et al. Effect of low-level carbon monoxide exposure on onset and duration of angina pectoris. Ann Intern Med 1973; 79: 46-50.

MMWR Deaths from Motor-vehicles-related unintentional carbon monoxide poisoning--Colorado 1996, New Mexico, 1980-1995, and United States, 1979-1992. 1996; 45: 1029-1032.

Morris RD, Naumova EN, Munasinghe RL. Ambient air pollution and hospitalization for congestive heart failure among elderly people in seven large US cities. Am J Public Health 1995; 85: 1361-1365.

Sheppard L, Levy D, Norris G, et al. Effects of ambient air pollution on nonelderly asthma hospital admissions in Seattle, Washington, 1987-1994. Epidemiology 1999; 10: 23-30.

Sheps DS, Herbst MC, Hinderliter SAL, et al. Production of arrhythmias by elevated carboxyhemoglobin in patients with coronary artery disease. Ann Intern Med 1990; 113: 343-351.

Thom SR, Ohnishi ST, Fisher D, et al. Pulmonary vascular stress from carbon monoxide. Toxicol Appl Pharmacol 1999; 154: 12-19.

Vellopoulou AV, Ashmore MR. Personal exposures to carbon monoxide in the city of Athens: I. Commuters' exposures. Environ Int 1998; 24: 713-720.

# CHAPTER 14.

# HEALTH EFFECTS OF HAZARDOUS AIR POLLUTANTS (HAPs)

When the 1990 Amendments to the CAA were passed, Congress called upon EPA to regulate 189 HAPs (also known as Air Toxics) as described in Chapter 4. Since it was not feasible to attempt to do controlled or epidemiologic studies of this many air pollutants, regulation required sources to keep the emission levels as long as possible using Maximum Available Control Technology (MACT). Congress did expect that EPA would be able to estimate the risk to public health from these compounds eventually. However data on outdoor concentrations of HAPs are not available for many communities. One study did attempt to estimate the public health risk from these so called Air Toxics (Woodruft et al, 1998). Approximately 10% of all census tracts had estimated concentrations of one or more carcinogenic HAPs greater than a 1-in-10,000 risk level. That survey found that background concentrations for eight of these HAPs exceed the no observed health effects level (NOEL, described in Ch 17). These eight compounds are listed in Table 14-1. The conclusion of this study was that HAPs may pose a potential public health problem.

*Table 14-1. Hazardous air pollutants that exceed the NOELs. Adapted from Woodruft et al, 1998.*

*Bis(2-ethylhexyl) phthalate*
*Benzene*
*Carbon Tetrachloride*
*Chloroform*
*Ethylene dibromide*
*Ethylene dichloride*
*Formaldehyde*
*Methyl chloride*

## HEALTH EFFECTS OF "AIR TOXICS"

Many of the HAPs are occupational pollutants. Also many have been categorized as carcinogens of suspected carcinogens. Since neither occupational pollutants nor carcinogens are covered in this text, there is only a short description of some representative studies.

## CONTROLLED HUMAN STUDIES

Several studies have attempted to recreate in the lab a chemical exposure similar to those implicated with "sick building syndrome". For instance, Koren and associates (1992) exposed 14 healthy subjects to a mixture of VOCs (25 mg/m$^3$ total hydrocarbon) for 4 hours (see Table 15-4 for the list). The investigators evaluated effects by using nasal lavage to detect upper airway inflammation. There was a statistically significant increase in neutrophil influx into the nasal passages after VOC exposure compared with air as shown in Table 14-2.

*Table 14-2. Concentration of total neutrophils (x $10^4$) seen in nasal lavage fluid after either air of VOC exposure.*

| *Time* | *Air* | *VOC* |
|---|---|---|
| *First baseline* | *24* | *87* |
| *Second baseline* | *21* | *21* |
| *Immediately post exposure* | *18* | *30* |
| *18 hours post exposure* | *15* | *108* |

*Adapted from Koren et al, 1992.*

A controlled study conducted at the University of Washington using the same VOC mixture as Koren and associates was designed to detect decrements in lung function or increases in bronchial airway or nasal inflammation (Pappas et al, unpublished data). In this study subjects were exposed to air, 25 mg/m$^3$ VOC mixture, or 50 mg/m$^3$ VOC mixture in a random, cross-over design where each subject received each treatment. The subjects were adults without asthma. Following VOC exposures there was a dose-related increase in lower respiratory symptoms. The symptoms were not associated with significant decrements in lung function or signs of inflammation in the subjects as a group. However, the subjects who were atopic did show a significant decrement in $FEF_{25-75}$ following exposure to the 25 mg/m$^3$ mixture.

Many of the complaints including in the “sick building syndrome” are non-respiratory including neurobehavioral symptoms. Otto and associates (1992) exposed a group of 66 healthy young, male adult subjects to the classic VOC mixture for 4 hours. The exposure protocol included one and one half hrs of clean air. 30 minute build up of VOC concentrations and then two hours and 15 minutes of VOC exposure. The assessment battery included 13 neurobehavioral tests. The subjects complained of fatigue and confusion, however the scores on the tests were not significantly different from scores obtained after the clean air control

Another assessment tool for determining adverse effects of VOC exposure is tests of eye irritation. In one study, 63 subjects were exposed in a chamber study to n-decane, one of the VOC used in the classic mixture (Table 15-4). The exposure was associated with increased tearing and increased numbers of conjunctival neutrophils indicating an inflammatory response (Kjaergaard et al, 1989).

## EPIDEMIOLOGICAL STUDIES

Kanawha Co, WV is one of the largest manufacturing centers of chemicals in the US. A research project investigated respiratory and irritant symptoms in all children grades 3-5 attending 74 elementary schools (Ware et al, 1993). Concentrations of 15 VOCs were measured at each school. Exposures were characterized by school location in relation to chemical plants, by the sum of concentrations of specific industrial processes, and by the concentrations of petroleum-related compounds. The valley had higher concentrations of these compounds that the surrounding area at higher elevations. Children enrolled in schools within the valley had higher rates of doctor-diagnosed asthma OR 1.27 (1.09-1.48); and a higher score for five chronic lower respiratory symptoms OR= 1.08 (1.02-1.14). Characteristics of children in the study are shown in Table 14-3. The cumulative incidence of selected health outcomes is given in Table 14-4. As shown in Table 14-4, physician-diagnosed asthma, the presence of chronic lower respiratory (defined as the presence of chronic cough, wheeze or physician-diagnosed asthma) and eye irritation were significantly more prevalent in children living in the valley than in children living out of the valley. These data suggest a chronic, adverse effect of the combined exposures.

Another epidemiologic study recruited 88 adult subjects (ages 20-45) in Uppsala, Sweden for an investigation of the effects of VOCs, formaldehyde and $CO_2$ in indoor atmospheres. Room temperature, air humidity, respirable dust, $CO_2$, VOCs, formaldehyde, and house dust mites were measured in the homes of the subjects. Asthma symptoms were most closely related to presence of dust mites and visible signs of dampness or mold. However, the investigators found a significant association between nocturnal breathlessness and $CO_2$, VOCs, and formaldehyde.

Based on a review of the occupational medicine literature, it is not surprising that adverse respiratory effects are associated with VOC exposure in human subjects. In the workplace, exposure to toluene diisocyanate has been found to cause occupational asthma. Isocyanates are the most common cause of occupational asthma. It is expected that research in the next decade will identify the health effects of many of the HAPs.

*Table 14-3. Characteristics of children by the proximity to industry of 74 elementary schools, Kanawha County, West Virginia, 1988. FROM: Ware et al, 1992 with permission.*

| Sample characteristic | In valley | | Out of valley | |
|---|---|---|---|---|
| | Near | Far | Near | Far |
| No. of schools | 16 | 26 | 14 | 18 |
| No. of children | 1,566 | 2,768 | 1,122 | 2,136 |
| % of children | 20.6 | 36.5 | 14.8 | 28.1 |
| Male sex | 49.4 | 51.1 | 52.3 | 50.3 |
| Race | | | | |
| White | 90.2 | 84.6 | 91.9 | 97.5 |
| Black | 8.0 | 13.2 | 4.9 | 1.1 |
| Other | 1.8 | 2.3 | 3.2 | 1.4. |
| Passive smoke exposure | 50.7 | 52.6 | 47.2 | 48.4 |
| Socioeconomic status | | | | |
| High | 12.1 | 14.0 | 23.3 | 13.9 |
| Medium high | 21.7 | 19.4 | 19.8 | 17.5 |
| Medium low | 30.5 | 30.0 | 24.9 | 25.7 |
| Low | 35.8 | 36.6 | 32.1 | 42.9 |
| Gas cooking stove in the home | 40.0 | 38.3 | 35.0 | 30.8 |
| Exposure to mold, mildew, or water damage | 24.7 | 22.4 | 25.5 | 21.3 |

* All data shown are percentages unless otherwise indicated.

*Table 14-4. Adjusted* cumulative incidence (%) of selected health outcomes by the proximity to industry of 74 elementary schools, Kanawha County, West Virginia, 1988. FROM: Ware et al, 1992 with permission.*

| Health outcome | Total | In valley | | Out of valley | | Trend test $p$ |
|---|---|---|---|---|---|---|
| | | Near | Far | Near | Far | |
| Chronic lower respiratory symptoms | | | | | | |
| Chronic cough†,‡ | 8.7 | 9.7 | 8.6 | 7.4 | 8.7 | 0.25 |
| Chronic phlegm† | 7.7 | 8.2 | 8.5 | 6.4 | 7.1 | 0.06 |
| Bronchitis† | 8.2 | 8.0 | 8.9 | 8.4 | 7.5 | 0.37 |
| Persistent wheezing†,‡ | 14.0 | 14.1 | 15.1 | 12.2 | 12.7 | 0.04 |
| Attacks of shortness of breath with wheezing† | 11.1 | 11.6 | 11.9 | 10.6 | 9.8 | 0.03 |
| Physician's diagnosis of asthma‡ | 10.1 | 10.4 | 11.2 | 9.5 | 8.5 | <0.01 |
| Composite indicators | | | | | | |
| Lower respiratory symptoms | 26.1 | 26.6 | 27.3 | 23.8 | 25.1 | 0.09 |
| Chronic lower respiratory response | 20.9 | 21.6 | 22.1 | 18.6 | 19.7 | 0.04 |
| Acute symptoms | | | | | | |
| Eyes | 7.3 | 7.8 | 8.1 | 7.3 | 5.7 | <0.01 |
| Nose | 25.4 | 24.8 | 24.9 | 24.2 | 26.6 | 0.23 |
| Throat | 21.2 | 21.1 | 20.5 | 19.9 | 22.6 | 0.22 |
| Lower respiratory symptoms | 9.9 | 11.2 | 9.4 | 8.7 | 10.1 | 0.43 |

* Adjusted for parental smoking and familial socioeconomic status.
† Health outcome included in lower respiratory symptoms composite indicator.
‡ Health outcome included in chronic lower respiratory response composite indicator.

## REFERENCES

Kjaergaard S, Molhave L, Pedersen OF. Human reactions to indoor air pollutants: n-decane. Environ Int 1989; 15: 473-482.

Koren HS, Graham DE, Devlin RB. Exposure of humans to a volatile organic mixture III. Inflammatory response. Arch Environ Health 1992; 47: 39-44.

Norback D, Bjornsson E, Janson C, et al. Asthmatic symptoms and volatile organic compounds, formaldehyde, and carbon dioxide in dwellings. Occup Environ Med 1995; 52: 388-395.

Otto DA, Hudnell HK, House DE, et al. Exposure of humans to a volatile organic mixture. I. Behavioral assessment. Arch Environ Health 1992; 47: 23-30.

Pappas GP, Pariser RJ, Henderson WR, et al. The respiratory effects of volatile organic compounds. Unpublished data.

Ware JH, Spengler JD, Neas LM, et al. Respiratory and irritant health effects of ambient volatile organic compounds. The Kanawha County health study. Am J Epidemiol 1993; 137: 1287-1301.

Woodruff TL, Axelrad DA, Caldwell J, et al. Public health implications of 1990 Air Toxics concentrations across the United States. Environ Health Perspect 1998; 106: 245-251.

# CHAPTER 15.

# HEALTH EFFECTS OF INDOOR AIR POLLUTION

When the Clean Air Act was passed in 1970, there was an assumption that controlling concentrations of serious wide-spread outdoor air pollutants such as the criteria pollutants and the hazardous air pollutants (HAPs) would protect the public from adverse effects of air pollution. Another early assumption was that blue collar workers were the exposed occupational population and that white collar workers were not subjected to harmful air pollutants on the job. However, sometime in the 1980s it became apparent that much of the exposure to the public from air pollutants occurred in the home and to workers from environments in office buildings. Workplaces with traditional air pollution exposures such as steel mills, saw mills, refineries and chemical plants had been targeted for regulation by the Occupational Health and Safely Act (OSHA) in the early 1970s. Indoor office buildings had not been included under OHSA and private homes contained totally unregulated atmospheres. In the late 1970s in conjunction with various problems with imported oil from OPEC countries, the US instigated recommendations for energy conservation which included constructing "tight" buildings. Buildings were built without operable windows and air exchange rates were reduced to save energy. Soon scientists and the public were hearing about the "sick building syndrome". Of course it was not the building but rather the human occupants who were "sick". All across the US clusters of office workers complained of a list of diverse and often subjective symptoms such as those listed in Table 15-1.

*Table 15-1. Symptoms associated with indoor office environments termed "sick building syndrome"*

*Irritation of eyes, nose, upper airways, throat, and skin.*
*Annoyance from odors*
*Respiratory function decreases in nonasthmatics including wheezing, cough, chest tightness, and shortness of breath.*
*Neurological symptoms including nausea, dizziness, headache, loss of coordination, tiredness, and confucion*

*Immunological reactions including inflammatory reactions, inmediate and delayed hypersensitivity allergic reactions*
*Asthma (aggravation and possibly causation)*
*Cancer*
*Respiratory infections*
*Increased susceptibility to infections or adverse responses to chemical substances*

---

Currently indoor air pollution is ranked by the US EPA and the Centers for Disease Control and Prevention in the top five environmental risks. Not only are there many toxic substances found in indoor environments but the exposure duration is great since most people spend the majority of their times indoors. The amount of time spent indoors is always listed as greater than 90% and in some surveys greater than 95%. Also various populations such as infants, the elderly, and those with pre-existing respiratory diseases spend nearly all their time indoors. As society became aware that indoor air was just as likely to be associated with adverse health effects as outdoor air, scientific studies began to be conducted to evaluate the health effects of indoor air. The issue of indoor air pollution has been reviewed recently in a chapter by Lambert and Samet (1996) and in a book (Bardana and Montanaro, 1997).

## INDOOR AIR POLLUTION SOURCES

At this point, I will describe some of the air pollutants that have been detected in indoor air. Lists of indoor air pollutants vary, however lead, tracked-in soil, volatile organic compounds, hazardous household chemicals, tobacco smoke, $NO_2$, CO, $SO_2$, formaldehyde, pesticides, and airborne allergens appear on all lists. Further adding to the difficulty sorting out health effects of indoor air and instituting remediation is the fact that many of the indoor pollutants associated with respiratory symptoms are natural. That list includes molds, dust mites, cockroach allergen, cat and dog dander (see Table 15-2). Perhaps the major indoor air pollutant affecting respiratory health is environmental tobacco smoke (ETS). When the area of the health effects of air pollutants gained prominence (around 1970) it was assumed that studies with subjects who were non-smokers would control for the confounding effect of cigarette smoke. Gradually it was realized that many non-smokers are exposed to second-hand cigarette smoke (termed passive smoke or ETS, the term used in this book). This greatly complicates differentiating the effects of indoor air pollution (as well as outdoor air pollution) from the effects of tobacco smoke per se.

*Table 15-2. Common sources of indoor air pollution*

| *MAN-MADE* | |
|---|---|
| *Environmental tobacco smoke (ETS)* | *Cigarette, cigar, pipe smoking* |
| *Nitrogen Dioxide ($NO_2$)* | *Gas cooking stoves* |
| *Carbon Monoxide (CO)* | *Home-heating devices including fireplaces and wood burning stoves* |
| *Fine Particulate matter* | *ETS*<br>*Wood Stoves* |
| *Formaldehyde* | *Particle board cabinets, etc.*<br>*Foam insulation (especially mobile homes)* |
| *Organic compounds* | *Paints, solvents, glues, hobby materials* |
| *Pesticides* | *Flea bombs, tracked-in garden pesticides* |
| *Chloroform* | *Chlorinated water used in showers* |
| *Perchloroethylene* | *Freshly dry cleaned clothes* |
| *Benzene* | *Cigarette smoke; attached garage gasoline fumes* |
| *NATURAL:* | |
| *Dust mites* | *Throughout the house* |
| *Molds* | *Bathroom, houseplants, water damaged areas* |
| *Animal dander and saliva* | *Pets* |
| *Pollens* | *Seasonal* |
| *Chloroform* | *Showers and other uses of hot water* |

# EXPOSURE ASSESSMENT: MAN-MADE

## Environmental Tobacco Smoke (ETS)

Environmental tobacco smoke has been measured at concentrations as high as 100 μg/m$^3$ for a monthly average when there is more than one smoker in a home. Figure 15-1 gives the monthly average fine particle concentrations indoors for no smokers, 1 smoker, and greater than 1 smoker compared to outdoor particle levels (NAS, 1986).

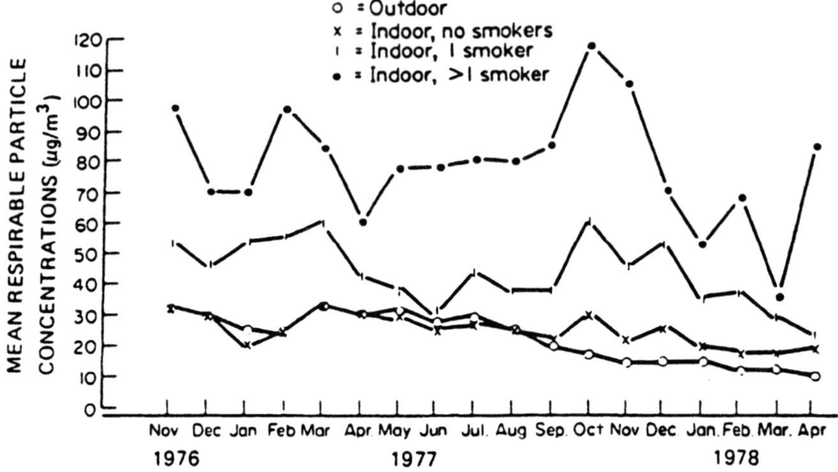


*Figure 15-1. Monthly average fine particle concentrations* indoors associated with *cigarette smoking. FROM: NAS, 1986.*

## Nitrogen Dioxide

Nitrogen dioxide concentrations can reach 1.5 to 2 ppm when a gas stove is lighted. In general, annual concentrations of indoor $NO_2$ are 8-15 ppb in homes without gas cooking stoves and 15-30 in homes with gas cooking stoves. These data have been summarized in the recent document, Air Quality Criteria for Oxides of Nitrogen (EPA, 1991). The health effects of indoor $NO_2$ exposure are described in Ch 12.

## Carbon Monoxide

Carbon monoxide (CO) can be present in any indoor spaces that have unvented or poorly vented combustion sources. Cases of CO poisoning have been reported from fire-place embers when the draft was closed prior to complete extinction of the fire. CO also can be emitted from a malfunctioning natural gas or oil furnace. Also CO concentrations become elevated inside a home if cars, lawn mowers, or snow mobiles are started and left running in an attached garage. The health effects of CO are discussed in Ch 13.

## Formaldehyde

Formaldehyde concentrations indoor can vary tremendously dependent on the sources of the gas and the tightness of the home. Because mobile homes tend to be built such that there is relatively little air transfer between inside and outside, and because these homes use materials indoors than can off gas formaldehyde, concentrations indoor mobile homes are often higher, in both a relative and absolute sense, than in other homes. There is a consensus that exposures of the public to formaldehyde should be less than 0.1 ppm.

## Organic Compounds

There are many sources of organic compounds within a home. These same sources are present in office buildings, however offices often have other sources as well from copy machines, carbonless copy paper, and so forth. Very little information on indoor levels of organic compounds exists. Common sources of indoor organic compounds are solvents, gasoline (especially if there is an attached garage), printed materials, dry-cleaned clothes, caulks, adhesives, particle board, moth crystals, stains, paints, polishes, and room fresheners. As discussed below, concentrations of organic compounds are often 5 to 10 times higher indoors than outdoors. A newly developed passive sampler being tested at the University of Texas at Houston offers hope that exposure assessment data for common VOCs will be available soon and assist in interpretations of these effects.

## Chloroform

Chloroform can be an indoor air pollutant. When chlorinated water is heated and used in the home in showers or washing machines or the kitchen sink, some of the chlorine is emitted into the breathing zone as chloroform.

## Pesticides

Pesticides sold for household use such as impregnated strips, foggers or “bombs”, which are classified as semi volatile organic compounds, include other chemicals as well (EPA, 1994). These pesticides and others tracked into the home from use outdoors may cause harm. The exposure can be from vapors in the room or from dusts in carpets and flooring. Small children who play on the floor are most vulnerable. Symptoms of pesticide exposure include dizziness, headache, muscular weakness, and nausea.

## Traffic Density

Traffic density in the streets near by residences may be a surrogate for the amount of exposure to vehicle exhaust air pollution that individuals experience. Since the fine particles emitted by vehicles readily penetrate indoors, traffic emissions become an indoor air problem. One study made repeated measurements of personal, indoor, and outdoor $PM_{10}$ for 37 nonsmoking adult subjects aged 50-70 (Janssen et al, 1998). A regression of unadjusted data gave a median R values of 0.50 for the correlation between personal and indoor/outdoor measurements. However when adjustment was made for ETS exposure over the past 24 hours, the correlation increased to 0.71. Adjusting for distance from a busy street further improved the correlation. In the final analysis the major difference between personal and outdoor concentration of $PM_{10}$ could be attributed to exposure to ETS, living along a busy road, and time spent in a vehicle. This study emphasizes the importance of understanding the activity patterns and behavior of subjects in exposure assessment studies, a point that was discussed in Ch 5.

## Benzene

Benzene can be present indoors if there is a cigarette smoker in the home. Also benzene can penetrate the home from gasoline fumes from a recently driven car in the case of an attached garage. Indoor benzene concentrations are a problem in Alaska since it is common to start the engine of the vehicle and allow it to warm up for many minutes. This activity provides a substantial exposure to benzene for residents. Benzene is an interesting example of mismatch between emission data and exposure data. Although 83% of benzene emissions in the US are from motor vehicles only 20% of the total exposure of individuals are estimated to come from this source. On the other hand, although cigarettes are responsible for less than 1% of the total emissions of benzene, the combination of benzene from active and passive smoking totals 61% of the estimated exposure to the average US residents (see Figure 15-2).

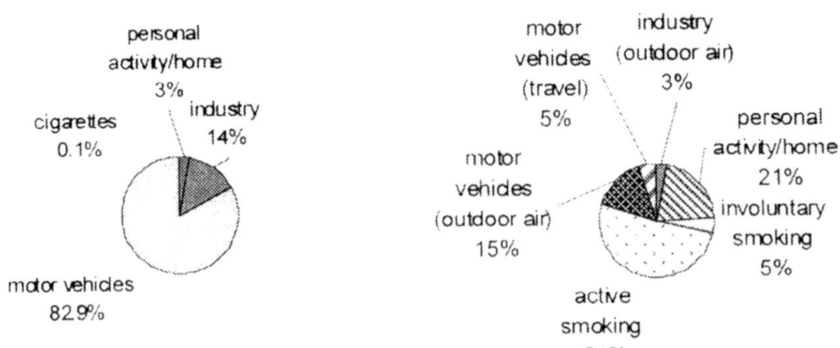


*Figure 15-2. A comparison of the sources of benzene emissions and the sites of benzene exposures for the US population.*

## TEAM STUDY

One of the most comprehensive studies designed to determine the exposure of individuals to indoor air pollutants within their homes was the Total Exposure Assessment Measurement (TEAM) study (Wallace et al, 1987) sponsored by US EPA. This study measured exposures in over 800 individuals of 20 volatile organic compounds (VOCs) and in a smaller group measured exposure to common pesticides. Two consecutive 12 hour air samples were gathered from all subjects. Also two 12 hour air samples were collected from the backyard of each subject. Finally each subject gave a sample of exhaled breath for chemical analysis. The individuals were selected to be representative of the US population. People were enrolled in Bayonne and Elizabeth, NJ, Greensboro, NC, Los Angeles, CA, and Baltimore, MD. Some rural sites were selected in North Dakota. Both personal and outdoor measurements were made so it was possible to apportion the sources. Personal samplers were worn by the subject for the entire 12 hour period. Personal samples were consistently higher than outdoor concentrations, sometimes as much as 10 times higher. The study was able to identify certain activities that were associated with high values during the 12 hour period. These were going to the gas station, dry cleaners, or smoking or being around a smoker. Also certain occupations were associated with higher exposure values. These were working with chemicals, plastics, or paints. Homes with smokers have significantly higher concentrations of benzene and styrene indoors than homes without smokers. Table 15-3 gives of list of 11 chemicals detected in exhaled breath of subjects in the TEAM study.

*Table 15-3. Concentrations (weighted means in $\mu g/m^3$) in exhaled breath from subjects in New Jersey during Fall, 1981.*

| *Chemical* | *Personal air* | *Outdoor air* | *Breath* |
|---|---|---|---|
| *1,1,1-trichloroethane* | *17* | *4.6* | *6.6* |
| *Benzene* | *16* | *7.2* | *12* |
| *m,p-Xylene* | *16* | *9.0* | *6.4* |
| *Carbon tetrachloride* | *1.5* | *0.87* | *0.69* |
| *Trichloroethylene* | *2.4* | *1.4* | *0.88* |
| *Tetrachloroethylene* | *7.4* | *3.1* | *6.8* |
| *Styrene* | *1.9* | *0.66* | *0.79* |
| *p-Dichlororbenzene* | *3.6* | *1.0* | *1.3* |
| *Ethylbenzene* | *7.1* | *3.0* | *2.9* |
| *o-Xylene* | *5.4* | *3.0* | *2.2* |
| *Chloroform* | *3.2* | *0.63* | *1.8* |

*FROM: Wallace et al, 1987 with permission.*

More recently, and more within the theme of this book, a similar study was conducted measuring exposure to Particle Total Exposure Assessment Measurement (PTEAM). In this study, particle concentrations were measured by personal monitors, indoor monitors and outdoor monitors (Clayton et al, 1993). The subjects were 178 nonsmoking individuals living in Riverside, CA. The personal sampler measured $PM_{10}$; the indoor and outdoor monitors measured both $PM_{10}$ and $PM_{2.5}$. Some of the filters from the monitors were analyzed by X-ray fluorescence (XRF) to determine chemical species making up the PM mass. As seen in the TEAM study, the personal monitors had higher concentrations than either the indoor or outdoor monitors. This was interpreted to imply that individuals are exposed to PM during activities that are not captured with the fixed monitors. Such activities are riding in vehicles, house cleaning, cooking, and perhaps being near a smoker. In the PTEAM study almost 25% of the $PM_{10}$ 24 hour averages were above the NAAQS 24 hour standard for $PM_{10}$. Daytime values were higher than nighttime values.

## INDOOR/OUTDOOR RATIOS

Numerous studies have shown that there is a good correlation between the level of outdoor and indoor fine particles. This correlation can easily be documented in a home without a smoker (since cigarette smoke is a major source of indoor fine particles). One study measured the concentration of indoor and outdoor fine particles simultaneously using a light scattering instrument (nephelometer) in nine homes during the heating season in Seattle. Each home was sampled for an average of 18 days. The average indoor/outdoor ratio of hourly values for all homes was 0.98. It was concluded that fine particles from outdoor sources such as residential wood burning are major source of indoor particles. The type of filtering system in the home affected the ratios (Anuszewski et al, 1998). Figure 15-3 shows the close tracking of outdoor and indoor fine particles during nighttime using simultaneous measurements indoors and outdoors (Timothy Larson, personal communication)

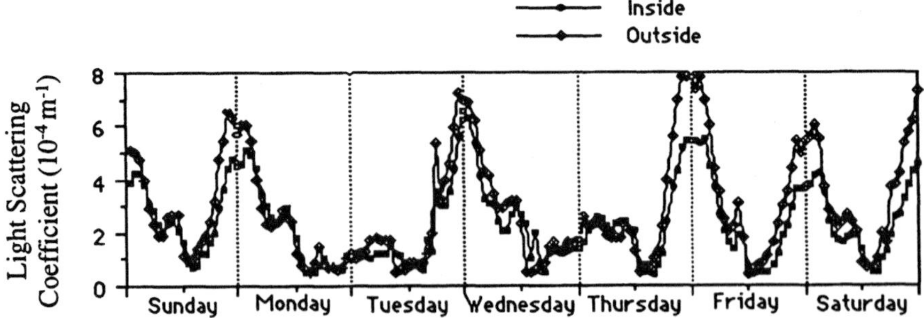


*Figure 15-3. Simultaneous indoor and outdoor measurements of fine particles by nephelometry during one winter time week in Seattle. FROM: Larson et al, 199.*

# HEALTH EFFECTS OF INDOOR AIR POLLUTION

## Environmental Tobacco Smoke (ETS)

ETS is a risk factor for the aggravation and development of asthma in children. The concentration of respirable particles inside a home is greatly influenced by the numbers of smokers. Compared to a home with no smokers particles are increased 2-fold with one smoker and 3-fold with two smokers (Figure 15-2). A recent summary by the US EPA(1992) documented the effects of ETS on acute respiratory illnesses in children including increased cough, wheezing, and phlegm production, acute and chronic middle ear diseases, and asthma aggravation. Chronic middle ear infections also are 20% more frequent in young children who are exposed to ETS. The relative risks for developing respiratory illness or symptoms that is associated with daily exposure to ETS are given in Figure 15-4. In general, prevalence rates of asthma were higher in children whose parents had both smoked during the child's lifetime. Murray and Morrison (1986) published findings that suggest that exposure to ETS in infancy is a risk factor for the development of asthma. In that study, the 24 children who lived with mothers who smoked had lower lung functions than the comparison children whose mothers did not smoke. Average $FEV_1$ values were 13% lower, average $FEF_{25-75}$ was 23% lower, and the exposed children had 43% more respiratory symptoms. These same investigators later published findings that suggest that children with dermatitis (a risk factor for the development of asthma) were more likely to become asthmatic if their mothers were smokers. This is a more serious charge than simply aggravating an existing condition. The causative agent in ETS is not known. ETS exposure is associated with 32,000 cases of cardiovascular disease per year in the US and up to 3000 cases of lung cancer in non-smokers.

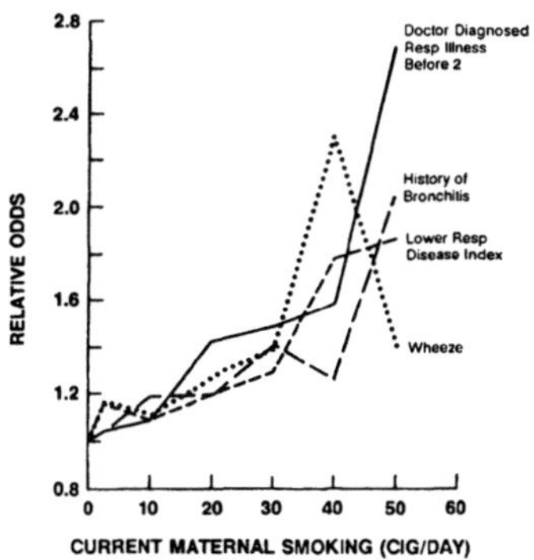


*Figure 15-4. Relative odds of respiratory illness of symptoms compared to maternal smoking frequency. FROM: NAS, 1986.*

## Formaldehyde

Animal studies of formaldehyde exposure have shown cell proliferation and death and nasal adenomas. Human studies indicate that, during exposure to formaldehyde, most of the gas (which is extremely water soluble) is taken up in the nasal passages. This may explain why no pulmonary function decrements have been seen in controlled studies of subjects with asthma at relatively high concentrations of the gas (3 ppm). It is also consistent with the reports of nasal and eye irritation from individuals exposed to formaldehyde in homes or work places. And it is consistent with the fact that a history of exposure to formaldehyde may be a risk factor for human nasal carcinoma.

## Organic Compounds

Many of the symptoms reported by individuals who claim that indoor air is causing health problems are headache, nausea, dry or itching eyes or noses, and nasal congestion. It seems highly possible that exposure to various organic compounds could cause these symptoms. On the other hand many other compounds or even stress from the workplace could cause these as well. Also these symptoms are difficult to quantify. There have been attempts to conduct controlled exposures to try to determine more objectively the human response to organic compounds. One controlled (although not blinded) study found that inhaling fumes from carbonless copy paper was associated with more upper airway symptoms than inhaling fumes from bond paper. Koren and associates conducted a controlled laboratory exposure to a mixture of 22 volatile organic compounds at 25 $mg/m^3$ in an exposure chamber (1992) [see Table 15-4]. That study found a statistically significant increase in neutrophils (a marker of inflammation) after VOC exposure compared to clean air. Likewise Danish investigators investigated whether exposure to vapors from organic solvents could cause lower airway effects in subjects with asthma (Harving et al, 1992). Eleven subjects with bronchial hyperresponsiveness were exposed to a mixture of solvents at 25 $mg/m^3$ for 90 minutes. $FEV_1$ decreased approximately 10% during the solvent exposures. A group at the University of Washington conducted a similar study evaluating controlled exposures to the same 22 VOCs for 4 hours in non-asthmatic subjects. Within the group those with a history of atopy showed dose-related increases in both upper and lower respiratory symptoms. No significant decrements in pulmonary function were seen. It is difficult to compare these studies using a mixture of many VOCs with the exposure of individuals who complain of symptoms in homes or offices.

*Table 15-4. Composition of the VOC Mixture used in controlled studies*

| *Chemical* | *Weight Ratio* |
|---|---|
| *N-Hexane* | *1* |
| *N-Nonane* | *1* |
| *N-Decane* | *1* |
| *N-Undecane* | *0.1* |
| *1-Octane* | *0.01* |
| *1-Decene* | *1* |
| *Cyclohexane* | *0.1* |
| *O-Xylene* | *10* |
| *Ethylbenzene* | *1* |
| *1,2,4 Trimethybenzene* | *0.1* |
| *N-Propylbenzene* | *0.1* |
| *2-Pinene* | *1* |
| *N-Pentanal* | *0.1* |
| *N-Hexanal* | *1* |
| *Isopropanol* | *0.1* |
| *N-Butanol* | *1* |
| *2-Butanone* | *0.1* |
| *3 Methyl - 2 Butanone* | *0.1* |
| *4 Methyl - 2 Pentanone* | *0.1* |
| *N-Butylacetate* | *10* |
| *Ethyloxyethylacetate* | *1* |

## Radon

Radon is an air pollutant found in some homes. It is a naturally occurring radioactive gas resulting from the decay of radium. It is a colorless, odorless gas that can enter indoor spaces from soil or rock formations under the structure. Since exposure to radon and its decay products in mines is associated with an increased risk of lung cancer, there is concern about indoor radon in homes. A number of case-control studies suggest that a lung cancer risk from indoor radon exposure (Axelson, 1995). In fact, radon is the second leading cause of lung cancer after smoking. Exposure to radon is enhanced in individuals who smoke since radon can attach to particles in cigarette smoke and be inspired into the lungs.

# NATURAL INDOOR AIR POLLUTANTS

Studies have measured concentrations of natural indoor air pollutants such as dust mites, molds, cat dander, and cockroach allergen inside the homes of children with asthma. High concentrations of these allergens were found in many homes. The presence of carpeting inside the home is one factor that increases the concentrations of both man-made and natural indoor pollutants. That fact led to the development of more efficient vacuum cleaners which can extract more of the dust from a carpet. Also vacuum cleaners are now available with filters which prevent dissemination of fine particles throughout the house. One vacuum comes equipped with a green light which lights up when the level of dust in the incoming air reaches a selected low value. High relative humidity indoors is another factor associated with increased levels of dust mites and molds, both proteins that thrive in humid air. It is recommended that the relative humidity indoors be kept below 50%, even a low as 45% if possible, to discourage growth of mites and molds. A recent survey of homes in the UK found the geometric mean concentration of dust mite allergen (Der p 1) to be 1.9 μg/g of dust. In that sample approximately 25% of the living room floor and bedroom mattresses has dust mite concentrations over 10 μg/g. Household characteristics associated with Der p 1 concentrations were age of carpet, window condensation, vacuum type, and presence of a smoker. Wide seasonal variation in levels of dust mite allergen has been documented which is mainly caused by variation in humidity (Platts-Mills et al, 1987).

It has been known for years that certain indoor agents (listed in Table 15-2) aggravate asthma. Bierman published a recent summary of the state of the art of environmental control for asthma symptoms (1996). Table 15-5 shows a list of allergens identified in house dust in the US. A number of studies are evaluating whether environmental interventions can remove these agents or at least decrease their levels in the home and whether the intervention will be associated with a decrease in asthma symptoms.

## Dust Mites

The relationship between dust mites and asthma has been known for over 70 years. Thirty years ago an assay was developed to measure the allergen associated with dust mites, Dermatophagoides spp. The source of this allergen is dust mite feces that are present throughout a house. Since dust mites feed on human skin flakes, the allergen is concentrated on mattresses, in bedding, and on overstuffed furniture and carpets. Studies clearly show that house cleaning and encasement of mattresses reduces the level of dust mites in the home. Fewer studies have actually shown that these reduced levels are associated with fewer asthma symptoms. Studies that remove the child from the home and send them to a more dust mite free environment (for instance high altitudes where dust mites are infrequent) have shown decrements in asthma symptoms and improvement in peak flow values.

*Table 15-5. Identified allergens in house dust in the United States. From Bierman, 1996 with permission*

| Name | Antigen Identification | | Notes |
|---|---|---|---|
| **Acarids** | | | |
| Dermatophagoides | | | |
| *D. pteronyssinus* | *Der f I* | *Der p II* | Common in America |
| *D. farinae* | *Der f I* | *Der f II* | Common in America |
| *Euroglyphus maynei* | *Euro m I* | | Uncommon in America |
| *Blomia tropicalis* | | | Common along the Gulf Coast of North America |
| **Insects** | | | |
| Cockroaches | *Bla g I* | *Bla g II* | |
| *Blatella germanica* | | | |
| Others | | | |
| **Domestic animals** | | | |
| Cat | *Fel d I* | | |
| Dog | *Can f I* | | |
| Others | | | |
| **Fungi** | | | |
| Multiple species including | | | |
| Penicillium | | | All three grow on surfaces or in rotting wood |
| Aspergillus | | | |
| Cladosporium | | | |
| **Others** | | | |
| Pollens from outside, wild mice and rats, other pets | | | |
| **Air pollutants** | | | |
| Cigarette smoke | | | |

## Cockroaches

In a recent study of inner city children with asthma, it was found that hospitalizations in the past year were linearly correlated with concentrations of cockroach allergen in the home. In that study, cockroach allergen was a greater predictor of hospitalizations than either dust mites or cat allergen (Rosenstreich et al, 1997). Figure 15-5 shows hospitalization frequencies compared with cockroach, dust mite and cat allergen exposure. These findings provide evidence that exposure to cockroach allergen has an important role in aggravation of asthma among children. The results suggest that attempts should be made to reduce the amount of cockroach allergen in the homes of children with asthma.

## Cats

Controlled exposures to cat allergen have shown that the allergen is capable of inducing symptoms in subjects with asthma who have a history of sensitivity to cats. Providing a cat free environment is very challenging as it has been shown that airborne cat antigen could be detected in a house 15-40 weeks after removal of the cat. Other studies show that cat allergen is present in homes and schools where no cat lives. It is assumed that the cat allergen travels on the clothes of cat owners and becomes airborne in many environments.

## Fungi and Molds

Fungi are among the most successful organisms on this planet as they have exploited every conceivable niche (Harriet Ammann, Personnel communication). They can thrive at both high and low temperatures. Molds are a form of fungi. Exposure to molds can be a source of respiratory aggravation. Molds are known to cause at least four adverse health conditions; infection, allergy, toxicity, and mucus membrane irritation. Thus molds are most often a problem when they are in a form that can be inhaled. However, some molds can cause dermatitis upon contact with the skin. The mold found indoors that has been associated with the most reports of aggravation of asthma is alternaria. Other molds found indoors known to affect persons with allergy and/or asthma are aspergillis, penicillium, and cladosporium. Recently a less common mold, stachybotrys, has been associated with some severe illness in children.

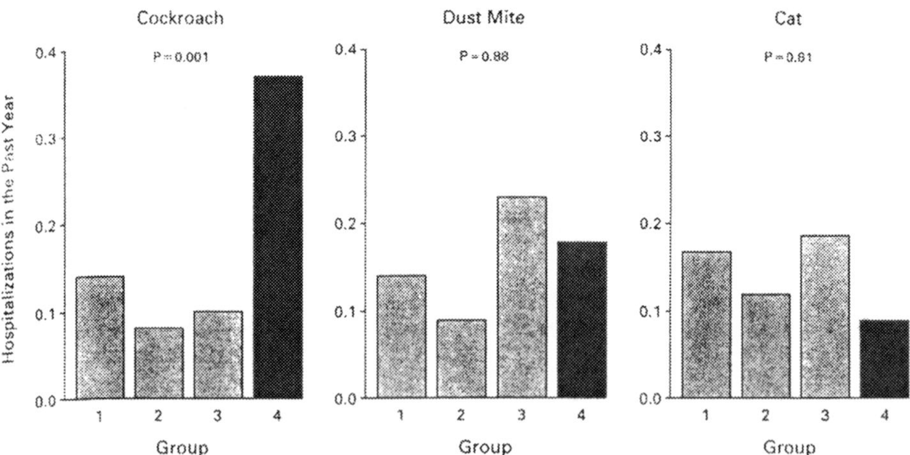


*Figure 15-5. Number of hospitalizations in the past year compared with the presence or absence of allergy to agents. Group 1 is children without allergies with low levels of exposure; group 2 is children without allergies with high levels of exposure; group 3 is children with allergies with low levels of exposure; group 4 is children with allergies with high levels of exposure. FROM: Rosenstreich et al, 1997 with permission.*

## CONTROL OF NATURAL INDOOR POLLUTANTS

Dust mite exposure to subjects with asthma can be reduced. Impenetrable mattress and pillow covers are recommended for mite-sensitive subjects. Also dust mite levels can be reduced by washing bedding in hot water (> 130°F). Dust levels in general can be reduced. First, removal of shoes before occupants enter the house is recommended. Also use of a track in mat to remove dirt from shoes in useful. Second, the use of carpeting that cannot be washed is discouraged since it traps dirt. Third, frequent use of a vacuum cleaner outfitted with rotor brushes and an effective filter can greatly reduce dust levels in the house. Regarding pets, it is recommended that they be avoided completely if a resident has asthma. If this is not acceptable the cat (or dog) should be kept outdoors or certainly not allowed in sleeping areas. Moisture in the home can be reduced by the use of bathroom and kitchen fans. As mentioned earlier, it is recommended that relative humidity be kept below 50%. Checking the home for water leaks and repairing them is recommended. All of these efforts will decrease humid conditions that are ideal for the growth of molds (and dust mites). Cockroach allergen level in the home can be reduced by keeping food and water sources covered and by caulking cracks under the sink and elsewhere in the kitchen and bathroom.

## SICK BUILDING SYNDROME

The quality of indoor air is a major concern for the health of workers. There is a wide range of pollutants in the indoor workplace which are not associated with the nature of the occupation, but rather are common to many indoor workplaces and offices. These are often the same pollutants as seen in homes, perhaps in higher concentrations. Printers, fax machines, and copiers emit specific air pollutants that can be irritating. The symptoms, listed in Table 15-1, have been reported by office workers in various situations. The most straightforward method for reducing indoor air pollution is to remove the source. If that cannot be achieved, it is recommended to dilute the pollutant by increased ventilation. Unfortunately building air intake vents are often located near a street and can introduce vehicle fumes indoors. Also, for some pollutants it is necessary to provide air cleaning measures such as air filters. In terms of ventilation, it is not enough to simply tell individuals to open the windows and get some clean air into an indoor space. In fact in most modern office building, the windows do not open. Air exchange rates in buildings should be sufficient to keep indoor carbon dioxide ($CO_2$) levels below 1000 ppm, or ideally below 700 ppm. (Outdoor $CO_2$ levels average approximately 350 ppm.) Temperature and humidity levels are factors in the comfort of indoor space. It is also generally recognized that psychological factors can influence an individual's perception of indoor air irritation as they can influence other disease processes. The sensory reaction to odor obviously serves as an indicator of air quality. This topic was summarized recently by Winneke (1992) who emphasized the role of detectability, thresholds, adaptation, and a pleasantness-unpleasantness scale. Winneke's studies have shown that environmental annoyance to different indoor air stressors are influenced by age, perceived health, and (to a lesser degree) gender. Since symptom ratings are subjective measures they are difficult to quantify. Lung function measurements, fingerpulse volumes, pupillary responses, and olfactory evoked brain potentials can be used as objective indicators of reactions to indoor air pollutants. However many complaints that individuals often report are subjective such as headache, lethargy; these effects are not directly measurable.

## CONCLUSION

This chapter has summarized potential exposures to indoor irritants and allergens. Since individuals spend the majority of their time indoors the risk of adverse effects from indoor air pollutants is greatly enhanced. There is now a movement among health professionals to reduce these exposures. One such program in Seattle is the Master Home Environmentalist (MHE) program. The MHE program was established to educate the public on environmental home pollutant issues with volunteer coaches who interact with home residents. The MHE uses a survey tool to assess homes for potential pollution problems and to develop a plan to reduce the presence of these pollutants in the home. A pilot study found that 31 out of 36 households who participated in the MHE program reported making at least one of the changes recommended by the coach. The MHE program is now being used in a study of the effectiveness of removal of pollutants on the status of childhood asthma.

## REFERENCES

Anuszewski J, Larson TV, Koenig JQ. Simultaneous indoor and outdoor particle light-scattering measurements at nine homes using a portable nephelometer. J Exp Analy Environ Epidemiol 1998; 8: 483-491.

Axelson O. Cancer risks from exposure to radon in homes. Environ Health Perspect 1995; 103 (Suppl 2): 37-44.

Bardana EJ, Jr, Montanaro A. Indoor air pollution and health. Marcel Dekker, Inc, New York, 1997.

Bierman CW. Environmental control of asthma. Immunol Allergy Clinics of N Am 1996; 16: 753-764/

Clayton CA, Perritt RL, Pellizzari ED, et al. Particle total exposure assessment methodology (PTEAM) study: Distributions of aerosol and elemental concentrations in personal, indoor, and outdoor air samples in a Southern California community. J Exp Analy Environ Epidemiol 1993; 3: 227-250.

EPA. Respiratory health effects of passive smoking; lung cancer and other disorders. EPA/600/6-90/006. December, 1992.

EPA. Indoor air pollution. An introduction for health professionals. US Goverment Printing Office 523-21/81322, 1994.

Harving H, Dahl R, Molhave L. Lung function and bronchial reactivity in asthmatics during exposure to volatile organic compounds. Am Rev Respir Dis 1992; 143: 751-754.

Janssen NAH, Hoek G, Brunekreef B, et al. Personal sampling of particles in adults: Relation among personal, indoor, and outdoor air concentrations. Am J Epidemiol 1998; 147: 537-547.

Koren HS, Graham DE, Devlin RS. Exposure of humans to a volatile organic mixture. III. Inflammatory response. Arch Environ Health 1992; 47: 39-44.

Lambert WE, Samet JM. Indoor Air Pollution IN: Harber P, Schenker MB, Balmes JR (eds). Occupational and Environmental Respiratory Disease. Mosby, St Louis, 1996. pp 784-807.

Murray AM, Morrison BJ. The effect of cigarette smoke from the mother on bronchial responsiveness and severity of symptoms in children with asthma. J Allergy Clin Immunol 1986: 77: 575-581.

NAS. Environmental tobacco smoke. Measuring exposures and assessing the health effects. National Research Council, Washington DC, 1986.

Platts-Mills TAE, Hayden ML, Chapman MD, Wilkins SR. Seasonal variation in dust mite and grass-pollen allergens in dust from the houses of patients with asthma. J Allergy Clin Immunol 1987; 79: 781-791.

Rosenstreich DL, Eggleston P, Kattan M, et al. The role of cockroach allergy and exposure to cockroach allergen in causing morbidity among inner-city children with asthma. N Engl J Med 1997; 336: 1356-1363.

Wallace LA, Pellizzari ED, Hartwell TD, et al. The TEAM study: Personal exposures to toxic substances in air, drinking water, and breath of 400 residents of New Jersey, North Carolina, and North Dakota. Environ Res 1987; 43: 290-307.

Winneke G. Structure and determinants of psychophysiological responses to odorant/irritant air pollution. Ann NY Acad Sci 1992; 641: 261-276.

# CHAPTER 16.

# INTERACTIONS BETWEEN CLIMATE CHANGE AND AIR POLLUTION

"The climate of Earth is changing. Climatologists are confident that over the past century, the global average surface temperature has increased by about half a degree Celsius. This warming is thought to be at least partly the result of human activities, such as the burning of fossil fuels and the clearing of forest for agriculture. As the global population grows and national economies expand, the global average temperature is expected to continue increasing by an additional 1.0 to 3.5° C by the year 2100. Climate change is one of the most important environmental issues facing humankind." The introductory statement is from a recent article on climate change (Melillo, 1999).

Climate change, which is also referred to as global warming, is thought to be driven by the accumulation of various green houses gases in air circulating near the surface of Earth. $CO_2$ is categorized as a green house gas because part of its action is to trap temperature in the troposphere thus warming the surface temperature. Per capita $CO_2$ emissions are shown in Figure 16-1. Another interesting graphic relating to carbon dioxide ($CO_2$) concentrations and temperature over the past century is shown in Figure 16-2.

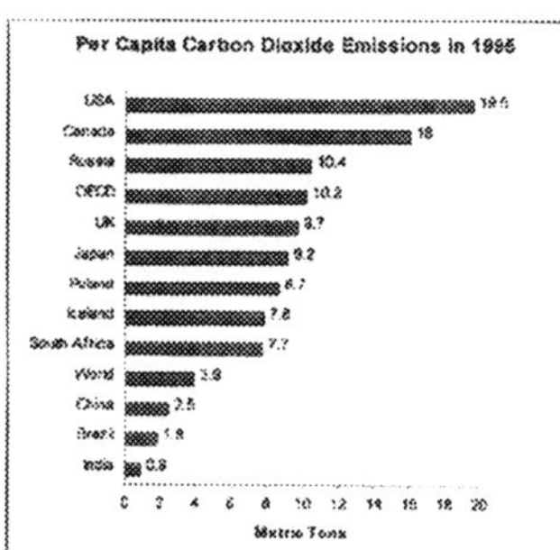


*Figure 16-1 Per Capita carbon dioxide emissions in 1995. Adapted from Raloff, 1998.*

$CO_2$ is not the only greenhouse gas but it is the most important contributor. Other greenhouse gases are methane, chlorofluorocarbons, nitrous oxide, and even carbon monoxide. The relative contribution of these various gases to the greenhouse effect is shown in Figure 16-3. It is not the purpose of this text to explore climate change per se. If the reader wants to delve more into the topic of climate change there is a comprehensive report published by the Intergovernmental Panel on Climate Change report (1995). Climate change is related to ambient air pollution because of the complex relationship between meteorology and air pollution. There is concern that climate change will affect air pollution in such a way as to be harmful to human health. A brief review of current thinking on the relationship between climate change and human health is presented, along with a summary of possible climate change effects on air pollution patterns.

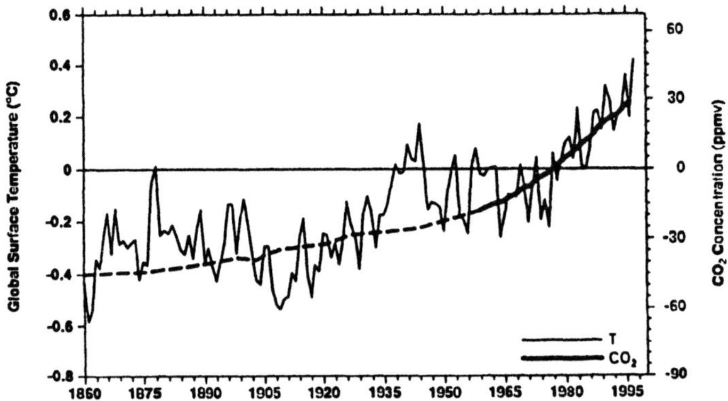


*Figure 16-2. Estimated changes in global-mean temperature (°C) (thin line) and $CO_2$ concentrations (thick dashed line from 1860 to 1995. FROM: Hurrell, 1998 with permission.*

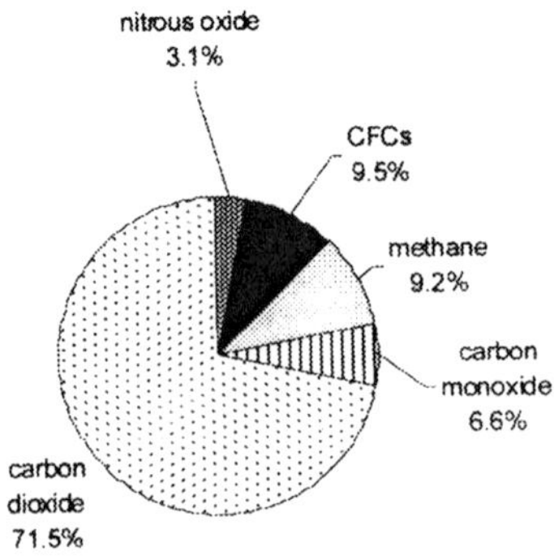


*Figure 16-3. Estimated contribution of various gases to global warming.*

## CLIMATE CHANGE AND HUMAN HEALTH: GENERAL

There appears to be a consensus that climate change will adversely affect health in at least four ways: an increase in skin cancer from the increase in ultraviolet radiation (UVB); an increase in cataracts and other adverse effects on the visual system from the increase in ultraviolet radiation; changes in immune system function as a result of temperature; and a change in the pattern of infectious diseases as vectors follow temperature increases. The first two examples--cancer and cataracts-- are direct effects of climate change. The third example, changes in the immune system, would be considered more systemic and indirect. Changes in infectious disease patterns also would be classified as indirect effects with various manifestations.

The first two health effects are dependent on changes in ultraviolet radiation. Ozone depletion and consequent increases in ultraviolet (UV) radiation that reach human populations is distinct from climate change. However, since the two phenomena are caused by some of the same gases, they are often discussed together.

There are two forms of UV light that reach the surface of the Earth: UVB which is in the light spectrum of 290-320 nanometers (nm) and UVC which is from 240-290 nm. Stratospheric ozone acts to filter the amount of UV radiation that is transmitted to Earth. There has been an increase in UV radiation reaching many geographical regions as of 1999 due to depletion of the stratospheric ozone layer. However, even without thinning of the ozone layer, it is projected that climate change alone will allow more UV radiation exposure to larger percentages of the population. Thus climate change and ozone depletion may act in an additive fashion to increase UV exposure. Ozone absorbs UV at wavelengths less than 315 nm. Thus ozone does not affect UVC radiation and but it does limit UVB radiation. Unfortunately for human populations UVB is the form of radiation associated with skin cancers and visual system effects. It is to our advantage to curtail depletion of stratospheric ozone. In 1974 scientists discovered that the ozone layer was being depleted by a family of chemicals known as chlorofluorocarbons (CFCs). In 1987, CFC levels in the stratosphere were increasing 5%/year. Due to the limitations placed on CFC use by the Montreal Treaty, CFC levels in the stratosphere are believed to be decreasing as of 1999. However there has been a major decline on stratospheric ozone over Antarctica associated with the CFC-induced depletion of ozone in the stratosphere.

## Skin Cancer

Table 16-1 shows the relationship between a projected 10% decrease in the ozone layer and increases in the percentage of UV depending on wavelength. There are three forms of skin cancer described in Table 16-2. Exposure to sunlight appears to be the major determinant of non-melanoma skin carcinoma. Tumors occur in light-exposed areas, increase with age, and are most frequent in light skinned individuals. On the other hand, melanomas have a complex relationship with UV. Only one type, lentigo maligna melanoma, has been directly related to sunlight. However epidemiologic evidence suggests other types also are associated. The incidence of skin cancer is generally increasing in Caucasians at 3-7% per year. EPA estimates that a 2.5% growth in CFC emissions will cause one million skin cancers per year (and 20,000 deaths). Although, the exact association between melanoma and UV radiation is still unknown, there is a linear relationship between incidence of melanoma and latitude (Figure 16-4).

*Table 16-1. Percentage increase in UV with a 10% depletion of stratospheric ozone*

*20% in UV at 305 nm*
*250% in UV at 290 nm*
*500% in UV at 287 nm*

---

*Table 16-2. Three major forms of skin cancer and their characteristics.*

*Non-melanoma skin cancers*

*1) Basal cell carcinoma*
*2) Squamous cell carcinoma*
*3) Melanoma*

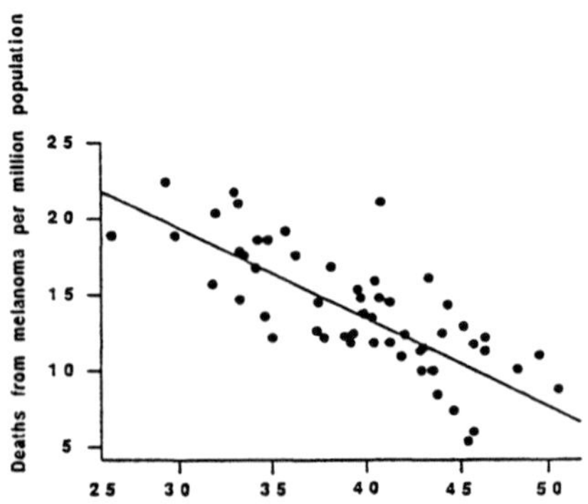

---

Deaths from melanoma per million population

25
20
15
10
5

25 30 35 40 45 50

*Figure 16-4. Geographical variation of melanoma mortality with latitude. FROM: Jones, 1987.*

## Cataracts

There are at least two adverse effects of UV radiation on the visual system; 1) cataracts and 2) retinal degeneration. UV radiation is considered a risk factor for development of cataracts. With the increasing life expectancies seen in many industrialized countries, more and more cataract surgery can be expected. Fortunately, present day removal of cataracts has become a relatively simple outpatient procedure and patients usually regain improved vision. On the other hand, retinal degeneration is –at the present time—a irreversible damage leading to blindness. Evidence of UV light exposure causing adverse effects on the retina is accumulating.

## Immune System

Effects of UVB radiation on the immune are not as well characterized as those on skin and eyes, but may turn out to be very important (Jeevan and Kripke, 1993). Some animal studies have found suggestive results that UVB is associated with immunosuppressive effects. The following effects are tentatively associated; loss of ability to mount an immune response to skin cancers, inhibition of delayed type hypersensitivity immune responses, and lack of ability to develop tolerance. At this time it is too early to make definite conclusions regarding the relationship between increases in UVB radiation and disruption of immune effects.

## Infectious Diseases

Increases in vector-borne diseases, which are highly correlated with temperature and humidity, are expected if tropical climates move northward. Effects can only be speculative, at present. An example is an introduction of malaria to North American as mosquitoes move northward from tropic climates. Mosquitoes are associated with the spread of yellow fever, dengue fever, and malaria. The Centers for Diseases Control and Prevention saw 43 cases of dengue fever from 18 states and the District of Columbia in 1996. Areas vulnerable to cholera may change with changing temperatures. With climate change, water ecology may change in ways that will enhance the spread of cholera. Also, world travel has changed the definition of disease prone areas. Cases of cholera, most likely transported from Latin America, have been seen in the US. Deforestation may change many vectors by interrupting their habitat. Table 16-3 lists major tropical diseases likely to spread with global warming.

*Table 16-3. Major tropical diseases that are likely to spread with climate change. FROM: Stone, 1995 with permission.*

MAJOR TROPICAL DISEASES LIKELY TO SPREAD WITH GLOBAL WARMING

| Disease | Vector | Population at risk (millions) | Prevalence of infection | Present distribution | Likelihood of altered distribution with warming" |
|---|---|---|---|---|---|
| Malaria | mosquito | 2100 | 270 million | (sub)tropics | +++ |
| Schistosomiasis | water snail | 600 | 200 million | (sub)tropics | ++ |
| Filariasis | mosquito | 900 | 90 million | (sub)tropics | + |
| Onchocerciasis (river blindness) | black fly | 90 | 18 million | Africa/Latin America | + |
| African trypanosomiasis (sleeping sickness) | tsetse fly | 50 | 25,000 new cases/year | tropical Africa | + |
| Dengue | mosquito | estimates unavailable | | tropics | ++ |
| Yellow fever | mosquito | estimates unavailable | | tropical South America and Africa | + |

SOURCES: WORLD HEALTH ORGANIZATION; *THE LANCET*
"As assessed by WHO: + = likely, ++ = very likely, +++ = highly likely

One of the possible effects of climate change may already be documented, that is increases in hantavirus infections. Hantavirus was diagnosed as the cause of a quickly debilitating flulike illness often leading to respiratory failure in the southwestern US in 1993 (Raloff, 1995). The hantavirus was traced to deer mice. The deer mouse population increased dramatically over the winter of 1992 that was uncommonly mild due to El Nino. It is assumed that as mild winters become more common through climate change, deer mice populations in many states will rise offering an increased risk of mice-human disease transmission. Thus hantavirus is seen as an emerging threat to human health in the US.

## CLIMATE CHANGE, TEMPERATURE, AND AIR POLLUTION

Temperature alone has its known health effects. Hot temperature causes heat stress and dehydration while cold temperature causes cold stress and hypothermia. Heat places stress on the thermoregulatory system and thus on the circulatory system. Individuals with pre-existing heart or lung disease can be extremely vulnerable to increases in temperature (Longstreth, 1991). For instance, in 1995 extreme high temperatures were associated with 700 excess deaths in Chicago. Both heat and cold adversely affect the cardiovascular system. Global temperatures in 1998 shattered the record high mark, making that year the warmest since at least 1860. It is not clear that all this excess temperature can be blamed on El Nino.

## Effects of Ozone

There is an association between ozone in ambient air and outdoor temperature. This relationship has been demonstrated experimentally in controlled laboratory studies and also with measurements of outdoor air. For example, in the Seattle area, the correlation between peak ozone measurements and temperature at the local airport is 0.80. Numerous studies have reported that years with high temperatures are associated with years with high summertime ozone levels. Figure 16-5 shows graphs of comparisons between maximum daily temperature and maximum daily ozone concentrations in four US cities. Based on these data, one would predict that climate change that leads to increased temperatures in urban areas will result in higher daily (and annual average) ozone concentrations. The expected increase in health effects from exposure to these increased ozone concentrations can be predicted from the data given in Ch 10 on the health effects of ozone. Briefly, one would expect more adverse pulmonary function effects in children playing outdoors during periods of elevated outdoor temperature, increased emergency room visits for asthma, and increased hospital admissions for asthma. We would also expect to see that higher temperatures in urban areas result in increased allergic reactions. This prediction is based on data that show that prior exposure to ozone lowers resistance to allergic stimuli. Outdoor sulfate concentrations may also be affected by temperature.

# OTHER POSSIBLE CONSEQUENCES OF CLIMATE CHANGE

Increases in allergic diseases may occur due to changes in molds and pollens caused by changes in vegetation patterns associated with climate change. This could lead to exacerbation of asthma and allergic rhinitis. There also is a possibility of increases in adverse developmental effects. Elevated body temperature can have an affect on fertility and neonatal development. There are suggestive data that show that perinatal mortality and preterm births occur more frequently during summertime.

Deforestation and desertification will both affect crops and if crop production decreases, there may be increases in malnutrition. Changes in water distribution can have deleterious effects. For instance, whether there will be sea level rise depends on whether more ocean water moves to glaciers (decreasing ocean levels) or whether glaciers disintegrate (increasing levels). Available water constitutes the major limitation to agricultural production. The current trend is toward less available water so any decrease would be added to an already serious problem. Models predicting effects on fisheries appear to be more complex and less known than those for agriculture. Episodic flooding may occur as warmer weather and increased winter precipitation are likely to increase frequency and severity of flooding. Floods can affect infectious disease vector patterns. There may be changes in river flow and water storage.

Reduction of snow pack would impact hydroelectric power as well as water supplies in the western US. A $3^{o}$ increase in temperature would lead to a 44% decrease in snow pack. $CO_2$ concentration, soil nutrients, temperature, and available water all impact the forests. Douglas fir might benefit by warming but hemlock would be disadvantaged. Warming could increase forest fires that then would influence pests and pathogens.

Increased population will impact all the potential disruptions mentioned above. There may also be direct adverse health effects from stress of crowding such as mental health disturbances. Finally a list of specific examples of health effects estimated from climate change is given in Table 16-4.

*Table 16-4. Specific examples of adverse health effects related to climate change.*

*1) Deadly cholera outbreak in Latin America in the mid 1990s*

*2) In Rwanda a $1^{o}$C increase in temp could lead to a 337% increase in malaria (other trouble areas are Costa Rica, Columbia, India, Kenya)*

*3) The hantavirus epidemic in southwestern US may have been indirectly tied to El Nino; 27 persons died. The relatively warm winter allowed an vastly increased population of deer mice, the carrier of hantavirus. The increased numbers of mice increased mouse-human contacts and thus the spread of disease. Hantavirus is carried in mouse urine. Sweeping up mouse droppings can aerosolize the protein and allow it to be inhaled by humans.*

*4) Increased summer heat waves would affect death rates in cities such as New York, Cairo, Shanghai, and Toronto. In 1995 there was a heat wave in Chicago where 700 excess deaths were recorded.*

---

Public health professionals need to identify the additive or synergistic contribution of humidity, air pollutants, and bioaerosols associated with adverse health effects of temperature. The American Conference of Governmental Industrial Hygienists (ACGIH) recently published a report on several aspects of environmental bioaerosols, including their assessment and control (1999). This report can be accessed on the internet at ACGIH.org.

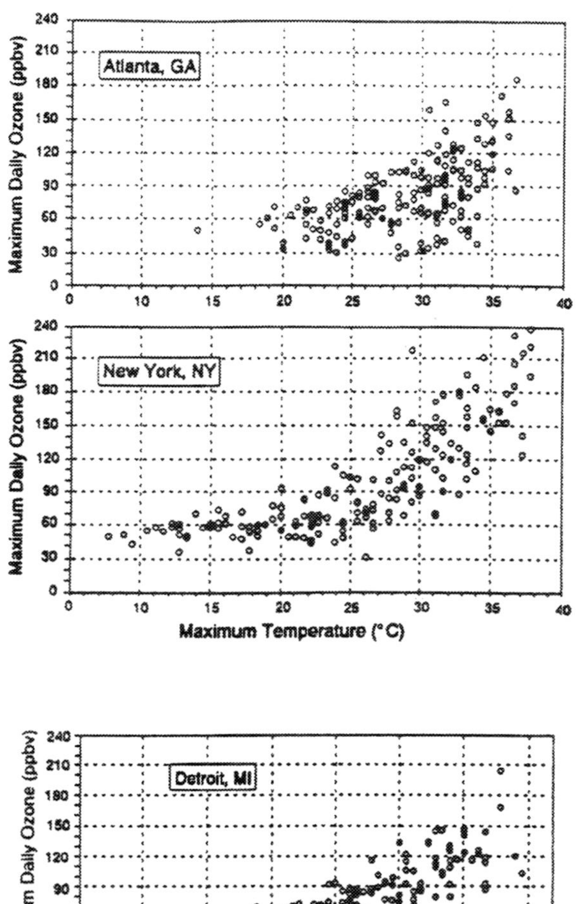

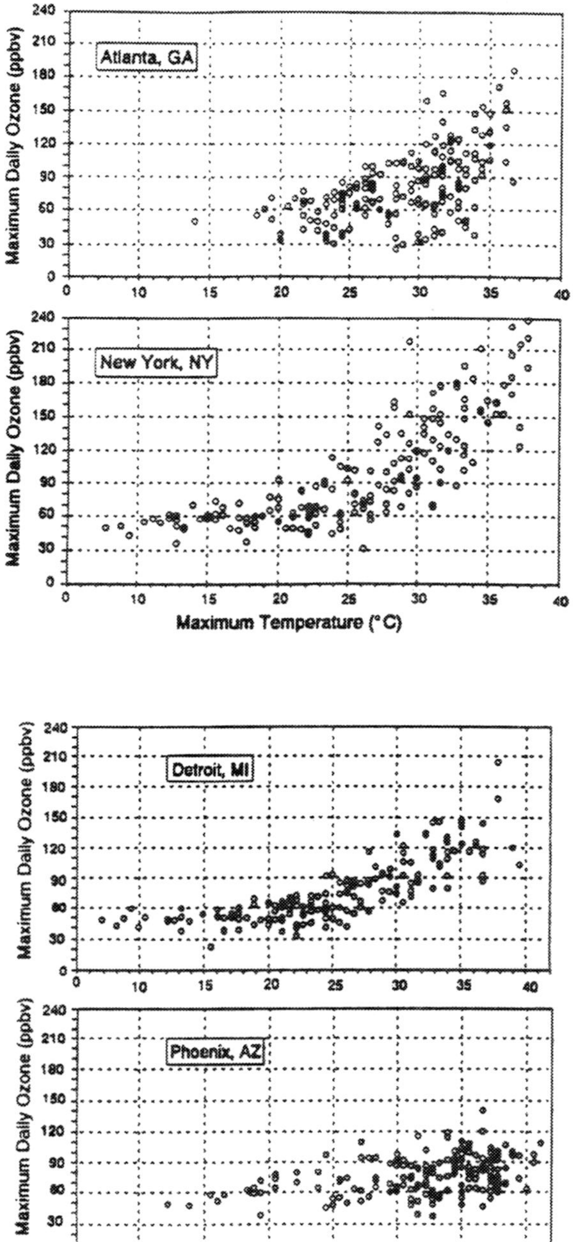

*Figure 16-5. A scatter plot of maximum daily ozone concentrations versus maximum daily temperature in four US cities. FROM: EPA, 1996.*

## CONCLUSIONS

It appears certain that clilmate change diven by man-made activitires is present and will be a factor for health effects in the future. Most likely it will increase many of the known health effects of the common air pollutants discussed in this text. The precise effect is unknown.

# REFERENCES

ACGIH. Bioaerosols: assessment and control. 1999.

EPA. Air quality criteria for ozone and related photochemical oxidants. EPA/600/P-03/004, Washington, DC. 1996.

Intergovernmental Panel on Climate Change (IPCC). Cambridge University Press, Cambridge, 1995.

Hurrell JW. Relationships among recent atmospheric circulation changes, global warming, and satellite temperatures. Sci Progress 1998; 81: 205-224.

Jeevan A, Kripke ML. Ozone depletion and the immune system. Lancet 1993; 342: 1159-1160.

Jones RR. Ozone depletion and cancer risk. Lancet 1987; 1: 443-446.

Longstreth J. Anticipated public health consequences of global climate change. Environ Health Perspect 1991; 96: 139-144.

Melillo JM. Warm, warm on the range. Science 1999; 283: 183-185.

Raloff J. How climate perturbations can plague us. Science News 1995; 148: 196-197.

Raloff J. Climate treaty talks mark some progress. Science News 1998; 154: 325

Stone R. If the mercury soars, so may health hazards. Science 1995; 267: 957-958.

# CHAPTER 17.

# RISK ASSESSMENT FOR AIR POLLUTANTS

This chapter discusses risk assessment, risk management, and risk communication in the field of health effects of air pollution. Risk is defined as a statistical probability that a harmful event will occur. Therefore, in the field of air pollution health effects risk pertains to the probability that exposure to a certain air pollutant will result in a given adverse physiiological effect in the defined population. A formal risk assessment protocol is divided into four parts: hazard identification, dose-response assessment, exposure assessment, and risk characterization (see Table 17-1).

*Table 17-1. Steps followed in a risk assessment protocol*

*Hazard identification*

*This is a first step that identifies potentially harmful agents to which persons may be exposed. This step attempts to establish a qualitative relationship between the agent and an effect. Next one asks several questions. Is the agent a potential human toxicant? Is an effect biologically plausible? Is there a potential exposed population and thus public health threat?*

*Dose-response assessment*

*The second step is to evaluate the dose-response relationship in terms of developing a quantitative relationship between dose of the agent and effects. Often the dose-response assessment has to estimate the dose to target tissue. Factors to consider are magnitude, duration, schedule and route of exposure, size, nature and class of the exposed population.*

*Exposure assessment*

*The third step is a process of quantitatively characterizing the relationship between the dose of an agent and the magnitude or incidence of an adverse health effect. The primary question addressed is whether people are exposed to the agent at levels that would have an adverse effect on health. This often requires extrapolation from high to low doses and from non-human species to humans.*

*Risk characterization*

*The final step integrates data from the other three components of the risk assessment process. The risk characterization summarizes the biological and statistical assumptions and identifies uncertainties. It also describes the strengths and limitations of the overall estimate of possible risk to humans (or other target populations).*

---

Risk assessment has become a major field for assisting in the determination of the safety of various situations, mainly the safety of the public from exposure to chemicals in water, soil, and air. Initially risk assessment was used for setting acceptable effects levels for carcinogens. More recently, risk assessment has been used to set standards to protect non-cancer endpoints as well. The process of setting the NAAQS actually is a form of risk assessment where the Hazard Identification process has already been performed (EPA has decided that the six criteria pollutants are potentially hazardous). In the criteria document written for each review of a criteria pollutant, authors assess the dose-response characteristics and the exposure assessment. Upon acceptance of the criteria document, EPA recommends the actual NAAQS for acceptance for the revised or re-affirmed standard.

## BASIC DEFINITIONS IN THE FIELD

This section describes some of the terminology used in the field of risk assessment (Lee, 1990). Maximum individual lifetime risk is used to denote the risk to the most exposed individual(s) for carcinogens and is expressed as a lifetime probability of contracting cancer. Acceptable daily intakes (ADI) are values for noncarcinogens derived through consensus by applying safety factors to LOELs or NOELs. LOEL stands for Lowest Observed Effect Level; NOEL stands for No Observed Effect Level. Sometimes these terms include Adverse effects and become LOAEL and NOAEL. These values were developed decades ago primarily for food contaminants, pesticides, drinking water, and ambient water. Reference Dose (RfD) is defined as an estimate, with an uncertainty of one order of magnitude or more, of a lifetime dose that is unlikely to pose significant risk to human health. The US EPA has established inhalation reference dose methodology (RfDi) to develop safe intake levels via inhalation. The RfD is calculated by dividing a NOAEL, obtained in dose response identification, by uncertainty factors (UF) and modifying factors (MF). In the event a NOAEL is not available, a lowest observed adverse effect level (LOAEL) is employed.

$$RfD = \frac{\underline{NOAEL} \text{ or } \underline{LOAEL}}{UF \times MF}$$

*Table 17-2. Uncertainty factors used in deriving reference doses and other standards.*
*<u>Standard Uncertainty Factors</u>*

*H = Human to sensitive human*

*Use a 10-fold factor when extrapolating from valid experimental results from studies using prolonged exposure to average healthy humans. This factor is intended to account for the variation in sensitivity among the members of the human population.*

*A = Animal to human*

*Use a 10-fold factor when extrapolating from valid results of long-term studies on experimental animals when results of studies of human exposure are not available or are inadequate. This factor is intended to account for the uncertainty in extrapolating animal data to the case of average healthy humans.*

*S = Subchronic to chronic*

*Use up to a 10-fold factor when extrapolating from less than chronic exposure results on experimental animals or humans when there are no useful long-term human data. This factor is intended to account for the uncertainty in extrapolating from less than chronic NOAELs to chronic NOAELs.*

*L = LOAEL to NOAEL*

*Use up to a 10-fold factor when deriving a RfD from a LOAEL, instead of a NOAEL. This factor is intended to account for the uncertainty in extrapolating from LOAELs to NOAELs.*

*D = Incomplete to complete data*

*Use up to a 10-fold factor when extrapolating from valid results in experimental animals when the data are "incomplete". This factor is intended to account for the inability of any single animal study to adequately address all possible adverse outcomes in humans.*

*<u>Modifying Factor</u>*

*Use professional judgment to determine another uncertainty factor (MF) that is $\leq 10$. The magnitude of the MF depends upon the professional assessment of scientific uncertainties of the study and data base not explicitly treated above; e.g., the number of animals tested. The default value for the MF is 1.*

---

*FROM: EPA, 1989*

## PERCEPTION OF RISK

Perceptions of risk vary among individuals. In general, people are more willing to accept risk if they derive some benefit than if they do not. For instance, most people do not avoid riding in automobiles although traveling in an automobile for just 300 miles is associated with a one in one million risk of an accident. On the other hand, people rate the risk from a nuclear plant accident very high although very few people have been harmed by such accidents in the US. Individual judgments of risk tend to be very subjective. People rate risks higher when they are associated with involuntary activities, they involve unfamiliar technologies, or when they appear unfair, as seen in Table 17-3.

*Table 17-3. Examples of risk-based priorities ranked into high, medium, and low risk.*

*High Risk*
- *Habitat alteration and destruction*
- *Species extinction*
- *Stratospheric ozone depletion*
- *Global climate change*

*Medium Risk*
- *Herbicides; pesticides*
- *Surface water pollution*
- *Acid deposition*
- *Airborne hazardous air pollutants (air toxics)*

*Low Risk*
- *Oil spills*
- *Groundwater pollution*
- *Radionuclides*
- *Acid runoff to surface waters*
- *Thermal pollution*

*Unranked Risks*
- *Ambient air pollutants*
- *Worker exposure to chemicals in industry and agriculture*
- *Indoor pollution*
- *Pollutants in drinking water*

---

*FROM: EPA, 1990*

## EXAMPLES OF RISK ASSESSMENTS FOR COMMON AIR POLLUTANTS

### $PM_{10}$

Using risk of death as an example, the consensus estimate is 1% increase in mortality for each 10 $\mu g/m^3$ increase in $PM_{10}$. For instance, if 24 hr $PM_{10}$ for one site was 30 $\mu g/m^3$ on average but increased to 60 $\mu g/m^3$ for a 2-3 day period, one would estimated 3% of the increase in deaths were attributable to $PM_{10}$.

Risk for an emergency department visit for asthma for each 10 $\mu g/m^3$ increase in $PM_{10}$ is 3.4%. With respect to risk of pulmonary function decrement, one study of elementary school children found an average 1.8 ml decrease in FVC for each $\mu g/m^3$ $PM_{2.5}$ in an community where 80-85% of the particulate matter came from residential wood burning (Koenig et al, 1993).

### Ozone

Risk of decrements in pulmonary function is used for this example. Lioy and co-workers (1985) reported a 0.55 ml decrement in FVC per 1.0 ppb of ozone in a study of children at a summer camp. If average ozone concentration is 80 ppb during the photochemical season in a given community, one would estimate an average 44 ml decrement in lung function. This would represent approximately a 5% decrement in a young child with a forced vital capacity (FVC) of 1.0 L.

### $NO_2$

Since an important suspected health effect of exposure to $NO_2$ is risk of increased respiratory illness, infectivity is used as the example of risk. Based on relevant studies, one would estimate a 40% increase in respiratory symptoms (cough, shortness of breath, wheeze, or bronchitis) in young children when annual $NO_2$ levels were increased by 15 ppb. One can also predict a risk of increased BHR associated with $NO_2$ exposure. It is estimated that BHR will increase in subjects with asthma at a concentration of approximately 100 ppb $NO_2$.

### $SO_2$ and $H_2SO_4$

The example of risk for these two sulfur oxide air pollutants is risk of increased BHR. Although the estimates are not well worked out, one study found an odds ratio of 1.62 for increased bronchial hyperresponsiveness when $SO_2$ increased 10 $\mu g/m^3$ (Seyseth et al, 1995). In screening for $SO_2$ sensitivity, a 10% reduction in $FEV_1$ after $SO_2$ exposure was the criteria for categorizing a subject as $SO_2$ sensitive. There also are data that allow one to estimate the risk of pulmonary function decrement. Decreases in FVC equal to 37 ml/$\mu mol/m^3$ $H^+$ were seen in a controlled laboratory study (Hanley et al, 1992), whereas a decrease in $FEV_1$ equal to 0.99ml/nmol/m$^3$ $H^+$ was seen in a community study (Studnicka et al, 1995).

## ETS

ETS is used as as example of an indoor air pollutant for which health risks can be precicted. EPA estimates that exposure to ETS from living with a smoker, results in up to 3000 deaths from lung cancer per year. The odds ratios for the development of asthma if the mother smokes was 2.1 (EPA, 1992).

The chapters on health effects of the individual pollutants have many more examples of data that could be used in a risk assessment. In fact, as mentioned earlier, the criteria document for each criteria pollutant that EPA prepares prior to the reconsideration of a NAAQS is basically a risk assessment for that pollutant. Table 17-4 gives a list of groups within the general population most likely to be affected adversely by air pollution due to specific host factors that define the group.

*Table 17-4. Groups most affected by air pollution*

| *High-Risk Group* | *Quantification in Population* | *Pollutants of Concern* |
|---|---|---|
| *Immature immune* | *Children below age 10-12* | *Respir system irritants* |
| *Deficient immune* | *Elderly people, progressive with age* | *Respir system irritants* |
| *Pregnancy* | *20 females per 50,000 females per year* | *CO lead* |
| *Vitamin C deficiency* | *10-13 of infants, children and adults of low income groups receive less than the RDA* | *Arsenic, cadmium CO, lead Chromium Pesticides Mercury Nitrates Ozone* |
| *Vitamin E deficiency* | *7% of general population* | *Cadmium, lead, ozone* |
| *Calcium deficiency* | *65% of children age 2 to 3 receive less than the recommended daily allowance* | *Lead* |

| | | |
|---|---|---|
| *Asthma* | *nearly 25,000,000 of general population* | *Respiratory irritants* |
| *Chronic respiratory* | *6,000,000--10,000,000 of general population* | *Resp disease irritants* |
| *Heart disease* | *15,000,000 of general population* | *CO, ozone $SO_2$* |
| *Smoking* | *Widespread* | *Cadmium, lead, polycyclic aromatic HCs radium 226* |

*Respiratory irritants: ozone, $NO_2$, sulfates, $SO_2$*
*RDA = recommended daily allowance,*
*HCs = hydrocarbons*
*Adapted from Calabrese, 1978.*

## RISK MANAGEMENT

Risk management is the process of managing the risk of a given pollutant. This involves deciding what to do about an assessed risk or groups of risks. With the criteria air pollutants, EPA has decided to set NAAQS and require that all communities in the country meet these standards (or at least have an active plan on how to meet the standards). For the hazardous air pollutants (HAPs) that are not criteria pollutants, EPA has decided to require that exposure to the population be controlled by requiring all emitters to use Maximum Achievable Control Technology (MACT). Many states have set reference standards for the HAPS. A diagram of the integration of risk assessment and risk management processes is shown in Figure 17-1.

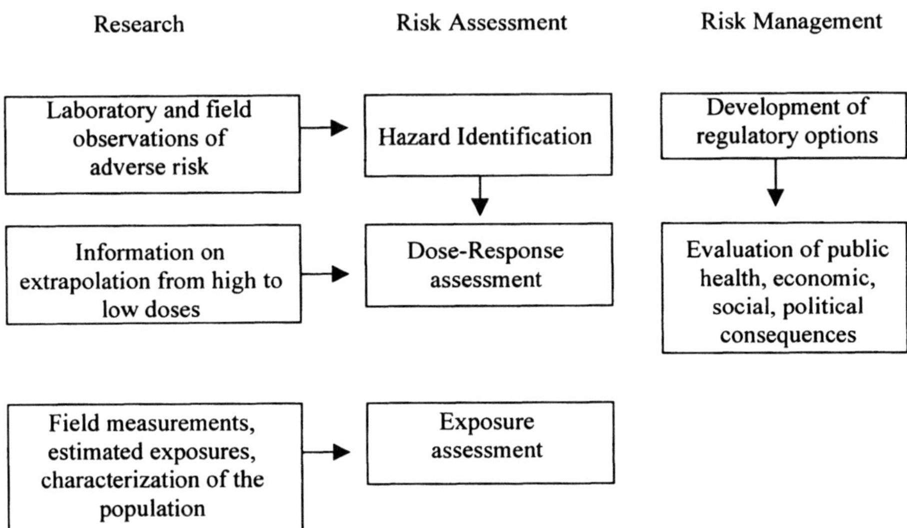


*Figure 17-1. Elements of research, risk assessment, and risk management. Adapted from NRC, 1983.*

Risk assessment should be used to set up a risk management protocol. Risk management traditionally has involved the development of regulatory options. This further involves evaluation of public health, economic, social, and political consequences of various regulatory options. Regulation based on risk assessment should be developed in parallel with a risk communication program. However in recent years there has been a movement to modify risk management from a regulatory driven process to a process that is driven by stakeholders. Stakeholders are defined as individuals who are exposed to the agent of concern, are emitters of the agent of concern, or have some other specific interest in the agent.

## RISK COMMUNICATION

Risk communication is probably the least well developed of the three processes. It refers to the process by which agencies and individuals attempt to communicate the risk of a given exposure to the exposed population or the general public. Many attempts at risk communication have been criticized on the basis that the audience did not understand the explanations. Also, because perceptions of risks often differ widely, a risk assessment communication has to be tailored carefully to the audience. See Table 17-3 for a ranking of some risks.

Effective risk communication requires first of all an effective characterization of risk and an understanding of the identity of the stakeholders involved. Risk communication requires description of the scientific risk assessment in non-technical terms so that the general public can understand the message. Risk communication should include clear messages about the nature, severity, and likelihood of risk (NRC, 1989). The assumptions, uncertainties, default factors and methods used to estimate the risk in question should be spelled out clearly. Risk communication is probably the most difficult aspect of the entire risk assessment process.

## STRENGTHENING THE FOUNDATON OF RISK ASSESSMENT

Plans for the future of risk assessment are to strengthen its scientific foundation by increasing the understanding of mechanisms and dose-response relationships, the identification of sensitive subpopulations, and evaluation of the relevance of animal models for estimating human risks. Also there is a need to develop novel and more reliable approaches to estimate human risks; collaborate with regulatory agencies on conducting risk assessment; and communicate findings to the public in an understandable and objective way. Some examples of strengthening the accuracy of risk assessment are 1) recent studies of susceptibility to carcinogens in cigarette smoke that found the frequency of at-risk genotypes vary significantly among Asians, Caucasians, and African-Americans and 2) studies that show individuals who carry the at-risk genotype for glutathione transferase μ, an enzyme that detoxifies constituents of cigarette smoke, suffer a 70% increased risk of bladder cancer. These examples both involve cancer, however it is expected that markers of susceptibility will be identified for noncancer risks in the near future.

Another issue involved in risk assessment is the issue of ethnic equality, also referred to as environmental justice. There is a belief, and some data, that show that low income, minority populations are more likely to live in proximity to Superfund sites, chemical waste facilities, and so forth. The fact that exposure to toxic agents is not randomly or uniformly distributed in the US but is instead often concentrated in minority and low-income communities, means that cultural, sociological, and psychological factors may determine how laboratory results are used in society.

A significant drawback to increasing the accuracy of risk assessments is the expense. Industry spends hundreds of billions of dollars to minimize risks to workers, to pay for health care, and to carry out litigation. For example, lawsuits surrounding Agent Orange exposure are estimated with a potential price tag of $500 billion.

What does the future hold? Some hope that the use of markers identified through molecular toxicology will hold promise for the future of human risk assessment. By sequencing specific genes (genetic biomarkers) in mice, scientists were able to detect induced liver cancer from background control tumors. It is hoped that application of genetic biomarkers for human effects will allow risk assessment to be more application.

# REFERENCES

Calabrese EJ. Methodological Approaches to Deriving Environmental and Occupational Health Standards, John Wiley and Sons, 1978, p. 55.

Dockery DW, Pope CAIII. Acute respiratory effects of particulate air pollution. Ann Rev Public Health 1994; 15: 107-132.

EPA. Interim methods for development of inhalation reference doses. EPA/600.8-88/066F. Research Triangle Park, 1989.

EPA. Reducing risks: Safety priorities and strategies for environment protection. Washington, DC. 1990.

EPA. Respiratory health effects of passive smoking: Lung cancer and other disorders. EPA/600/6-90/006F, December, 1992

Hanley QS, Koenig JQ, Larson TV. Response of young asthmatic patients to inhaled sulfuric acid. Am Rev Respir Med 1992; 145: 326-331.

Infante-Rivard C. Childhood asthma and indoor environmental risk factors. Am J Epidemiol 1993; 137: 834-844.

Koenig JQ, Larson TV, Hanley QS, et al. Pulmonary function changes in children associated with fine particulate matter. Environ Res 1993; 63: 26-38.

Lee SD. Risk assessment and risk management of noncriteria pollutants. Toxicol Indust Health 1990; 6:245-255.

Lioy PJ, Vollmuth TA, Lippmann M. Persistence of peak flow decrement in children following ozone exposures exceeding the National Ambient Air Quality Standard. J Air Pollut Control Assoc 1985; 35: 1068-1071.

Morris RD, Audet A-M, Angelillo IF, et al. Chlorination, chlorination by-products and cancer: A meta-analysis. Am J Public Health 1992; 82: 955-963.

Neas LM, Dockery DW, Ware JH, et al. Association of indoor nitrogen dioxide with respiratory symptoms and pulmonary function in children. Am J Epidemiol 1991; 134: 204-219.

NRC (National Research Council). Risk assessment in the federal government: Managing the process. National Academy Press, Washington, DC, 1983.

NRC (National Research Council) Improving risk communication. National Academy Press, Washington, DC, 1989.

Seyseth V, Kongerud J, Haarr D, et al. Relation of exposure to airway irritants in infancy to prevalence of bronchial hyperresponsiveness in schoolchildren. Lancet 1995; 345: 217-220.

Studnicka MJ, Frischer T, Meinert R, et al. Acidic particles and lung function in children. Am Rev Respir Crit Care Med 1995; 151: 423-430

# CHAPTER 18.

# STANDARD SETTING FOR AIR POLLUTANTS

The US has granted the EPA the authority to set standards for air pollutants in ambient air. The EPA was formed in 1970 and, since that time, has set standards for various environmental agents including air pollution. Up to that time, individual states had various requirements but there was no national effort to impose a single standard for clean air.

EPA developed a process for setting air quality standards that are called the National Ambient Air Quality Standards (NAAQS). The NAAQS have been referred to repeatedly through this text. This chapter describes the process for the setting of NAAQS. Other national standard exist for drinking water, workplace exposures, radiation exposure, food and drug quality, toxic substances, solid waste, pesticides, and so forth.

## DEFINITION OF AN ADVERSE EFFECT

In conjunction with setting health based standards, it is necessary to develop a consensus on what constitutes an adverse effect. This is not a simple process. What is an adverse health effect of environmental or occupational exposures? Obviously premature death is an adverse health effect. However there is not as much agreement on the degree of pulmonary function decrement that constitutes an adverse effect—10% 15% 20%? This is complicated by the fact that individuals do not all start at 100% of predicted lung function. Thus for a subject with asthma whose current $FEV_1$ is 75% of predicted normal $FEV_1$ a 10% decrement associated with ozone exposure carries a different degree of adversely than for a healthy young adult whose current $FEV_1$ is 105% of predicted. Is airway inflammation always a more serious effect than a reversible change in lung function? The American Thoracic Society (ATS) published a statement on "What constitutes an Adverse Health Effect of Air Pollution" in the 1970s. An ATS committee currently is writing a revised statement. Issues for the new statement are listed in Table 18-1.

*Table 18-1. Issues to be considered in a statement of adverse effects*

*Population versus individual effects*
*Ethics and equity*
*Economic costs*
*Susceptibility factors*
*Heterogeneity of perspectives*

---

Many scientists and public health professionals now agree that air pollution standards cannot protect all sensitive individuals. They certainly intend to set standards that will prevent exposures that interfere with daily life. However, ultimately it is society that must make the judgment where to draw the line. When the Clean Air Act was passed, language prohibited EPA from considering economic factors in setting the NAAQS. However economics did implicitly affect EPA decisions. As more and more research is conducted, and as assessment tools for health effects and air monitoring of air pollutants reach lower and lower detectable levels, it is clear that there is no threshold for the onset of adverse health effects that would be accurate for the entire population. Another complicating factor it that is appears there are different mechanisms for the different air pollutants. Thus using pulmonary function criteria as a definition of adverse effects may be valid for $SO_2$ exposures but may not be valid for adverse effects for $NO_2$ since these now appear to include subtle changes in immune system function. The basis of a standard should be supported by scientific consensus, there should be clear definitions of the nature and severity of adverse effects, and there should be descriptions of significant risk for both the general public and for sensitive sub-populations.

*Table 18-2. Dimensions of concern to be addressed in a statement of adverse effects*

*Use of biomarkers for exposure assessment*
*Quality of life estimations*
*Physiological impact*
*Symptoms ranking scales*
*Clinical outcomes*
*Mortality*
*Risk*

---

The relative importance of different adverse effects is often depicted using a pyramid such as the one shown in Figure 18-1. Minor effects are noted across the base of the pyramid with increasing adversity of effects moving up toward the ultimate adverse effect, death.

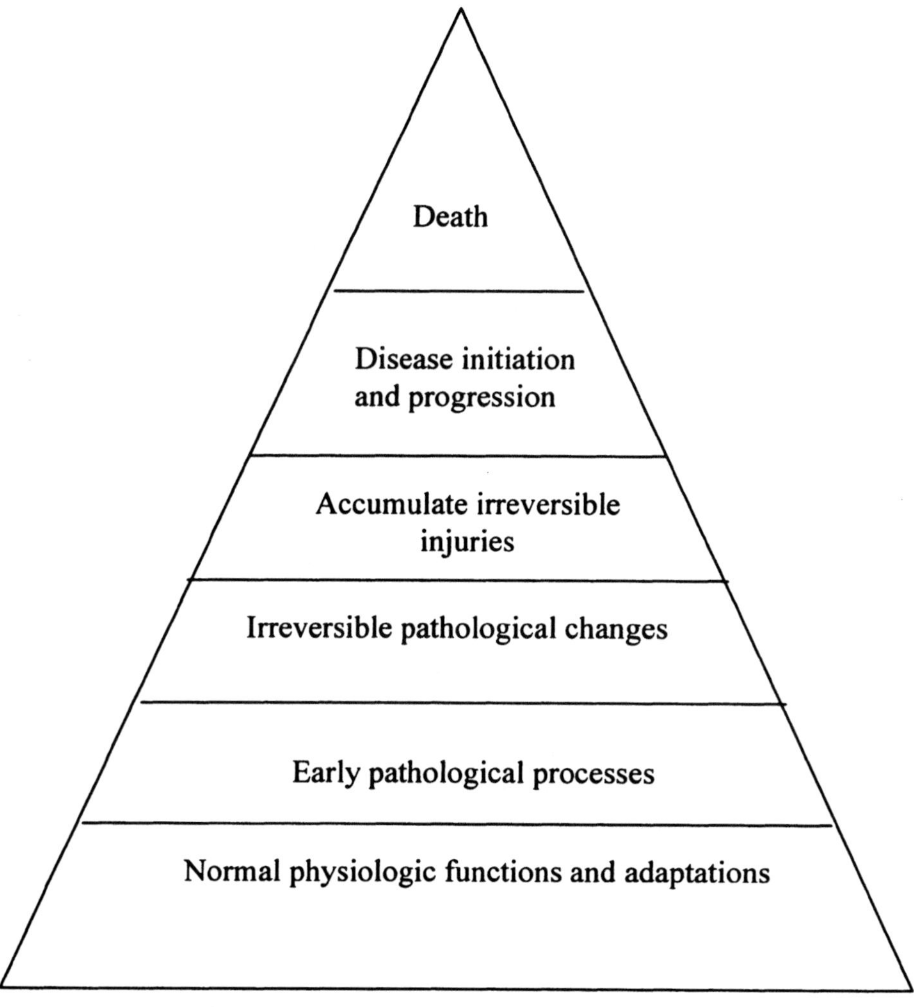


*Figure 18-1 Spectrum of health outcomes.*

All the issues that were discussed in Ch 17 on risk assessment are relevant for this discussion on standard setting. The two fields cannot be separated.

## STANDARD SETTING FOR NAAQS

By law, the NAAQS for each criteria pollutant is to be reconsidered every 5 years. Following is the process for setting NAAQS. The process is initiated with plans for developing a criteria document, staff plan, and ultimately, promulgation of a NAAQS. As part of this process, EPA develops the Criteria Document (CD) that is written by the National Center for Environmental Assessment (NCEA). There is a set schedule for the various required reviews. These include considerable internal EPA review, next a review committee of scientists is formed and they critique the CD. After several iterations of this process, the CD is ready for the initial review by the Clean Air Science Advisory Committee (CASAC) reviews each CD several times. Each time sends it back to NCEA for re-writing until it finally achieves approval by a majority of the committee. Also the public has an opportunity to provide testimony at scheduled CASAC meetings. After the public has had a chance to make written and oral comments and a majority of the CASAC members agree that the CD is adequate and complete it is accepted. Once it is accepted, another branch of EAP—the Office of Air Quality Planning and Standards (OAQPS)—writes what is called a staff paper to summarize the CD and make recommendations to the administrator. The recommendations list a range of concentrations and averaging times for the new NAAQS. After the Administrator has announced the CASAC recommendation for a new or revised NAAQS, there is a 90 day public comment period. Again, after the Administrator announced the concentration and averaging time selected, there is another public comment period before the NAAQS is listed in the Federal Register.

## LIMITATIONS OF NAAQS

How well does this system work? Not surprisingly there are a number of limitations and inadequacies to EPA standard setting process. Some of these are listed. Are National Ambient Air Quality Standards adequate to protect public health? What about regional differences in the pollutants? Should the NAAQS be for mixtures rather than single pollutants?

Regarding regional differences, both ozone and PM are heavily influenced by local conditions as has been described in other chapters. For instance, PM is dominated by sulfates on eastern seaboard and by wood smoke and other vegetative burning and in the west. Ozone precursors vary significantly from region to region. Many of the criteria pollutants are highly correlated: ozone, carbon monoxide, nitrogen dioxide, sulfur dioxide, and particulate matter are all correlated to some degree either depending again on the particular air shed (see Table 7-7).

The lengthy process involved in setting a standard prohibits a rapid adjustment of the NAAQS. At present there is over a year in preparation of the criteria document, prolonged time period for closure on the criteria document, and requirement for reviews by the Clean Air Science Advisory Board and the public. Should there be a restriction in the length of the criteria documents?

How can the most recent data will utilized? At present, data included in the CD have to have undergone peer review and acceptance for publication by a journal requireing peer reviw. Should unreviewed data be included in the CD? Should local air agencies be encouraged to set Guidelines for health effects standards which are different from the NAAQS ? At present, the state of California has standards for most of the criteria pollutants that are stricter than the NAAQS. Should EPA put out updates on health effects research during the 5 years between criteria documents? These could take the form of the MMWR published by CDC.

The NAAQS process involves writing an extremely comprehensive criteria document. The CAA specifies that EPA reviews each criteria pollutant on a 5-year schedule. Thus the agency is now writing the new PM. In spite of the fact that the new $PM_{2.5}$ standard was announced in the federal register on July 19, 1997, another document is in preparation. Does this make sense? How can people have confidence in this new standard when they know that a revision is already underway? Critics may conclude that the science and the policy are out of synch. However, it must be stressed that scientific studies proceed at the pace set by individual investigators with interest in health effects of air pollution. The science is not, and should not, be dictated by standard setting..

## NAAQS FOR PM: AN EXAMPLE

Prior to 1986, PM was regulated as total suspended particulate matter (TSP). However as a result of its regular update of the standard, EPA and CASAC decided to narrow the size range of PM and regulate just those particles which measure 10 micrometers or less in diameter. This category is referred to as $PM_{10}$; the decision to change to this designation was based on the fact that these diameter particles are the ones most likely to be respired into the lung (see Ch 10) ). At this time, $PM_{10}$ is one of the two pollutants (the other is ozone) which is of greatest concern for public health. According to the latest national emissions trends report from EPA almost 70 million Americans live in areas that do not meet these standards. The standards for both pollutants were recently reviewed by EPA. In fact, the American Lung Association sued EPA to speed up its review of the $PM_{10}$ standard based on recent studies showing an association between $PM_{10}$ and mortality. The courts gave EPA a deadline of 1997 by which to set a new standard or reaffirm the present standard that is 150 $\mu g/m^3$ averaged over a 24 hour period or 50 $\mu g/m3$ as an annual average. After considerable review and input from both the public and the Clean Air Science Advisory Committee, Carol Browner, the EPA administrator decleared a new health standard for PM. That new standard, also listed earlier, is for a smaller size fraction PM less than or equal to 2.5 micometers in diameter. The new standards are concentrations of 65 $\mu g/m^3$ $PM_{2.5}$ for a 24 hour average and 15 $\mu g/m^3$ PM2.5 for an annual standard. Communities will not have to demonstrate that they meet these standards until at least 2002 because they have to set up monitoring networds and determine which areas in each urban area requires a monitor.

Other agencies also set standards for some of the criteria pollutants for other purposes, for example for protection in the workplace. Table 18-3 lists health based standards set by other agencies that regulate environmental exposures.

*Table 18-3. Comparison of risk assessment and risk management estimates from various agencies.*

| *Estimate* | *Exposure scenario* | *Effect Level* | *Population* | *Population* |
|---|---|---|---|---|
| *ACGIH TLV-TWA* | *8 h/day 40 h/wk 40 yr* | *Impairment of health or freedom from irritation Narcosis Nuisance Stress* | *Healthy worker* | *Industrial Experimental human and animal* |
| *NIOSH REL* | *10 h/day 40 h/wk Working lifetime* | *Impairment of health or functional capacity and technical feasibility* | *Healthy worker* | *Medical Biological Chemical Trade* |
| *OSHA PEL* | *8 h/day 40 h/wk 45 yr* | *Impairment of health or functional capacity and technical feasibility* | *Healthy worker* | *Medical Biological Chemical Trade* |
| *EPA Rfc* | *24 h/day 70 yr* | *NOAEL* | *General population including susceptible* | *Occupational. Experimental human and animal* |
| *ATSDR MRL* | *24 h/day 70 yr* | *NOAEL* | *General population including susceptible* | *Occupational. Experimental humanand animal* |

*NAS = National Academy of Sciences*

*NIOSH = National Institute of Occupational Safety and Health*
*OSHA = Occupational Safety and Health Administration*
*ATSDR = Agency for Toxic Substance Disease Registry*
*TLV-TWA = Threshold limit values-time weighted average, the standard used by industry*
*REL =*
*PEL= Permissible exposure limit*
*RfC = Reference concentration*
*MRL =*

The populations served by ACGIH, NIOSH, and OSHA standards is working people. These standards are knowingly set to allow higher concentrations of pollutants than would be allosed for the public based on the assumption that workers are healthy. Due to this fact, the standards often do not protect workers with asthma or other chronic respiratory diseases. It can be seen that there is a range of concentrations and averaging times that currently are used in the US as standards for the protection of public health. EPA and ATSDR standards are set to protect the public at large.

# CHAPTER 19.

# CONCLUSION AND FUTURE DIRECTIONS

It is an exciting time for the field of the health effects of air pollution. Every month one hears of new, creative research designed to help understand the associations between ambient outdoor pollutants and human health.

The next decade should result in research findings that will improve our knowledge of the risk to health from common air pollutants. PM research is greatly on the rise due to action by both the National Research Council (NRC) and by EPA. The NRC has published a research agenda for the coming 13 years. EPA has just funded five centers in the US devoted to the study of the association between PM and health. Many of the gaps in our knowledge about the associations between PM and health were mentioned in Ch 10. To reiterate, we need to know much more about the chemical composition of PM. EPA is funding speciation sites around the country that will analyze PM filters for various chemical constituents. These data can be used by epidemiologists and toxicologists to ask more precise research questions. It is likely that we will know much more about the mechanisms of the effects of PM in the next 5 to 10 years. Also, cardiovascular endpoints recently have been added to the spectrum of PM health effects. More research will elucidate that relationship.

Two areas of future research into the health effect of ozone are worth emphasizing. One is the recent research showing an association between mortality and ozone exposures in a large cohort study. The coming decade will tell us whether this result is reproducible. The other area of interest is the examination of the effects of long term exposures to ozone (and other pollutants, for that matter). A study is planned to evaluate life time pollutant exposure on asthma outcomes in children enrolled in a multi-city study. The children will be followed for at least 10 years; symptom, medication, lung function, and unscheduled physician care will be available.

As mentioned in Ch 12, there is an increase in interest concerning $NO_2$ exposures. Studies that need to be carried out are those that can evaluate the role of $NO_2$ in common respiratory infections.

Early in 1999 there was a report of low birth weight in infants being associated with CO concentrations during the years of 1989-1993 in Los Angeles. Also there were two recent articles in one of the premier journals for reports of air pollution health effects. One article found an association with increased pulse rate and PM in the Salt Lake City area. This type of research begins to explain the adverse cardiovascular effects seen in time series studies. The other article presented data that show an association between chronic exposure to PM and mortality in non smokers. The result was only significant for males indicating that some other factors, perhaps occupation, plays a role. However, these findings suggest that gender differences in susceptibility to air pollution may exist. This issue has not be studied to any extent but has the possibility of great importance.

What lies ahead for regulation of ambient air pollutants? Some epidemiologic studies are finding associations with health outcomes and several air pollutants, rather than being able to identify a single pollutant. Will we be able to find single pollutant effects from exposure to the air pollution 'soup' that we breathe? Will EPA decide to regulate mixtures of ambient pollutants rather than single pollutants as they do now? California is setting the stage for new control technologies for air pollution. The state has set a deadline for reduction of sulfur content in diesel fuel sold in California. The state also has categorized diesel as a cancer-causing agent. Gasoline refineries are experimenting with new formulations of gasoline that will change the pollutant mix in our air. Manufactures are developing fuel cell technology to get cleaner emitting vehicles on the road. Fuel cells that burn hydrogen emit almost no air pollution. Electric cars have been suggested as a solution to the air pollution problem, however a sufficiently light battery has not yet been developed. The other problem with electric cars is the need to recharge the battery frequently.

In conclusion, it appears that the next decade will see many advances in our knowledge of the health effects of air pollution. One can only guess the content of a book similar to this written in the year 2009.

# INDEX

Printed in the United Kingdom
by Lightning Source UK Ltd.
135144UK00007B/113/A